CONCEPTS OF EVOLUTION

Everett C. Olson
Jane Ann Robinson

University of California at Los Angeles

Charles E. Merrill Publishing Company
A Bell & Howell Company
Columbus, Ohio 43216

Published by
Charles E. Merrill Publishing Company
A Bell & Howell Company
Columbus, Ohio 43216

Illustrations by Kathryn Bolles

NATURAL HISTORY MUSEUM
OF LOS
library
ANGELES COUNTY

Copyright ©, 1975, by Bell & Howell Company. All rights reserved. No part of this book may be reproduced in any form, electronic or mechanical, including photocopy, recording, or any information storage and retrieval system, without permission in writing from the publisher.

International Standard Book Number: 0-675-08739-2
Library of Congress Catalog Card Number: 74-19542
1 2 3 4 5 6 7 8 9 10 — 82 81 80 79 78 77 76 75

Printed in the United States of America

Preface

Concepts of Evolution was initially developed as a mimeographed text for a lower division college course for nonbiology majors. The aims of the text and the course were to provide a general introduction to biology in the context of evolution and to view organic evolution in its relationships to society and philosophy. The final version emerged after several years of teaching and extensive modification of the original mimeographed text as dictated by the interest, needs, and understanding of the students.

Students, ranging from freshman to seniors, from about twenty different segments of the University of California at Los Angeles have formed classes with which we have worked. The largest numbers have come from anthropology, geography, economics, mathematics, physics, English, and psychology. In spite of the varied backgrounds and interests, the current prominence of biology in everyday life and the common social and philosophical threads which are carried through the text have provided a common base of communication. We are deeply indebted to all of these students for their frank criticisms and evaluations of the text materials and for their many suggestions which have been incorporated in the final version. We hope that their aid may prove valuable for students to come who may read the book and, perhaps, find it stimulating.

Many of the subjects also have been explored in depth in graduate seminars and we are indebted to the participants of these as well for reading parts of the text, aiding in maturation of many of the concepts, and providing suggestions for additions and deletions. We wish in particular to acknowledge our debt to Robert Nelson, Sheri Gish, and Michael Bell who as graduate students in the Department of Biology at U.C.L.A. have worked closely with the course both as teachers and as critics. Deborah Duffield and Dorothy Oehler read parts of the manuscript and offered many valuable suggestions which have been incorporated in the final version.

The illustrations were done by Kathryn Bolles who was as much a collaborator in preparation of the book as she was the artist who depicted substantive materials and concepts in a sensitive and imaginative fashion. She was untiring in searches for the most appropriate materials for illustration and in her meticulous criticisms of the form and content of the manuscript. In order to catch the feel of the work, she attended the class upon which it was based for a full quarter. We are especially indebted to her for her many-faceted aid and take this opportunity to express our deep appreciation.

Everett C. Olson
Jane Ann Robinson

Contents

	Introduction	vii
1	General evolutionary concepts	1
2	The nature of life	13
3	The origin of life	29
4	The early evolution of life	51
5	The metazoans	73
6	Species, variation, and evolution	109
7	Examples of evolution in action	135
8	The evolution of social systems	171
9	The evolution of man	191
10	Cultural evolution - man	219
11	The impacts of evolutionary concepts	233

Introduction

In Biology, the principle of organic evolution by means of natural selection is the hub about which almost all other theories revolve. However, evolutionary theory has a much broader, although less obvious, influence on the world of today. Our understanding of the fabric of modern society and culture is incomplete without some knowledge of the roles that evolution plays in the day-to-day flow of events and ideas and in the development of societies' basic forms.

The climate for evolutionary thought was present as early as the seventeenth and eighteenth centuries, but the full force of this thinking was not felt until the publication of *On the origin of species* by Charles Darwin in 1859. Darwin demonstrated a simple and rationally acceptable mechanism for evolutionary processes, that of gradual change by means of natural selection. This was what initiated the rapid infusion of the concept of organic evolution into biology and eventually into society where, only rather recently, it has matured into a pervasive and largely subconscious way of thought underlying our societal, cultural, and ethical world views.

The study of organic evolution in biological curricula naturally is directed toward the more technical scientific aspects of the subject. As a rule, students learn about the mechanisms of inheritance and natural selection and from there depart into considerations of ecology, population genetics, and the history of life. This is a reasonable and proper approach for students who already have some background upon which they can call and who will eventually bring their specialized knowledge to bear on their own biological endeavors. Increasingly, in various texts, more attention is directed toward the history and meaning of the ideas behind evolution; however, these efforts tend to fall short of the needs of students both in biology and in other fields who desire an understanding of how evolutionary concepts relate to human events and spheres of knowledge beyond the scope of science. *Evolutionary Concepts* is designed to meet such a need, but without sacrificing a sound understanding of the biological basis for evolutionary theory.

The term *evolution* may be, and often is, applied to physical processes; but the phrase *theory of evolution* in its usual sense refers to organic evolution, the evolution of plants and animals. The idea of organic evolution arose initially from studies of living organisms in the attempts to explain their immense diversity, their interrelationships in dynamic systems, and their distribution around the world. Paradoxically, the self-same data can be used to argue for the creation of the earth and its life by an Intelligent Designer, God, who imparted order and harmony to all things. And although geology, paleontology, biogeography, and paleoecology have come to play important parts in the development of the idea that life has had a long changing history on earth, they too can be cited both by evolutionists and creationists alike to support their conflicting points of view.

The most pressing point in the early disputes between nineteenth century evolutionists and their opponents was not the broad problem of whether the changing events of history were real or not, but whether species were fixed entities unable to be transformed into other species in time. Breeding of plants and animals to emphasize particular traits had shown that species could vary, but the central question remained; could man or nature ever carry specialized breeding to the point where something truly new came into being? The future of human thought for years to come hinged on the response of the people of that time to Darwin's data and conclusions. If positive, the floodgate of evolutionary theory could be opened and its potential fully explored; if negative, the door to evolutionary thought could be emphatically and permanently closed. That the response was positive is a matter of historical record. Our investigation of the how and why of this acceptance will carry us into the mechanisms of change and the nature of species, the evidence for evolution.

Studies of species and the mechanisms by which they change can become quite involved and technical, but their meaning is enhanced if they can be put into the context of the evolution of the idea of evolution itself. Even more basically, it is important to comprehend something of what life is, how it came to be, and how the forms and functions of living things have changed over the eons of time since life began. So, in this book, we will first paint the events of organic history with a broad brush upon the canvas of ideas of life and change as viewed by our predecessors. Once this background is in hand, we will look at how species actually do originate and what biologists think of the role of present-day species—including man—in evolution today. Our approach may seem "backwards," since the material is not presented in the order followed by most texts. However, we feel we have chosen the easiest way to make progress toward an intuitive and rational understanding of evolution in all its manifestations.

Evolutionary change always involves the interactions of a great many individual organisms both within single species populations and among members of different species, and these are inextricably linked with changes in the physical environment. Taken together, these interrelationships point to a coevolution of plants, animals, and environment that forms the "big picture" of evolution. We cannot think only of changes in species and genera, but must also consider changes in complex organized systems. Within species, the most complex form of organization is the social system, and the evolution of such systems is the concern of a special branch of evolutionary study.

Complexes of cooperating individuals have arisen repeatedly among many different species of plants and animals, but the very highest levels of development are found among certain kinds of insects and in our own species, *Homo sapiens*. Once the principles of change at the individual, population, and ecosystem levels have been grasped, it is possible to look somewhat objectively at the intricate problems of social evolution. We are especially interested, as is natural, in the evolution of human societies, and it is in this very area that most of the conflicts about the reality of evolution arise. Insect societies, however, provide an opportunity to view an alternative type of system in which caste roles are rigidly defined and made mandatory by programmed development of individuals within the societies. Such closely knit social groups are sometimes called *superorganisms* and have served as models for fanciful societies in the future of man, as in Huxley's *Brave New World* and even the frightening *1984* of Orwell. We will look into the insect societies and use them as a contrasting backdrop for the study of the nature of man and his evolving social systems.

When we come to consider man, it becomes evident that something different has appeared on the evolutionary scene. Man is the first animal, so far as we know, to be fully conscious of himself. Just how this came to be is a question we will raise, but one which we will not and cannot answer fully. With the advent of consciousness, a "new" phenomenon—cultural evolution—begins to become significant. Cultures undergo gradual directional changes, that is, they evolve; but the mechanisms involved are quite apart from those operating in biological evolution. We will study the nature of cultural evolutionary processes just as we study those of biological change. In the course of the investigations, some highly pertinent and controversial problems will arise, and these will be discussed in chapter ten.

In the last chapter of the book we are in a position to complete the full circle of ideas begun in chapter one. First we asked what evolution was, and then we saw something of the history of life and how it could be explained. The explanation, the theory of organic evolution, is one of the great conceptual achievements of man; but ideas have a way of feeding back into the society that generated them and becoming the basic thought patterns by which its members live. The problems of all men become the problems of philosophers, and their writings, in turn, influence the society which generated them. Thus evolution has had an immense effect on everything we do, and even on the way we look at life. As we return to this point, now with a full background, we can see how evolutionary thinking has become entrenched in modern life and what some of its major consequences are and might be. In final analysis, a look at the past with an objective eye to the future is where any integrated study must come to rest.

1

General evolutionary concepts

Perspectives

Heraclitus (540–476 B.C.) conceived the universe and its life to be in a state of complete flux, this being the nature of things. At about the same time, Paramenides was attempting to prove that change was unreal, merely an illusion. Through the ages, from Plato, Aristotle, Spinoza, and Kant to Bergson and James, the argument on change and changelessness has gone on, and it continues today. Most of the arguments and *proofs* of change and nonchange have been of an abstract sort, dealing with such matters as time paradoxes and the nature of time, the changelessness of God, and whether a thing that has become different from what it was is, in fact, the same thing. Commonsense recognition of the flow of events and the nature of time usually have received far less attention.

Because we will be discussing evolution all through this book, we cannot and should not avoid investigating this matter of change. Evolutionary theory, as we will look at it most of the time, is a scientific theory; our starting point will be common sense, for that is where science begins. Perhaps we should not bother with the "philosophical nonsense" connected with evolutionary theory, but the philosophy is

CHANGE AS ILLUSION

PARAMENIDES (ca. 500 B.C.)
Change is only illusion; the *being*, comprising all existing things, is the ultimate reality.

PLATO (ca. 400 B.C.)
Through his senses, man perceives a changing world, but this is merely a shadow of true reality—the intellectual realm of *immutable essences*.

CHANGE AS REALITY

ARISTOTLE (ca. 350 B.C.)
Change is real, spontaneous, and continuous, and is directed toward a final cause.

HERACLITUS (ca. 500 B.C.)
All things are in ceaseless flux; only change, or *the becoming*, is real.

not easily disregarded. At the very least, if we are going to deal with only the "facts", we must have some concept of just what facts are. Furthermore, if we are going to look at the idea of evolution as it is now or as it has been at any time in the past, we must recognize that the health of an idea depends upon the intellectual climate of the time; that is, the idea depends on the philosophical outlook of the world in which it was nurtured. When there was an intellectual climate unfavorable to the idea of change, evolutionary ideas found hard sledding. The temper of the times tends to be a rather gross expression of the more erudite philosophy from which it came and draws from this philosophy the ground rules for which types of thinking are acceptable and which are not.

Today, we operate largely under the influence of a materialistic scientific concept of nature and man. We respect cause and effect, and we believe that what we hear, see, taste, and feel is real. We can perceive change in the physical world, and the change may be either of cyclical or linear character. These concepts represent common sense to us. The idea of evolution of the universe poses no intellectual problems. In such a climate, the idea of *organic* evolution also should thrive and, for the most part, it does.

The fact is, however, that many persons do find the idea of biological evolution, or at least parts of it, offensive. Many just do not believe it. Atomic theory, relativity, and quantum theory all are acceptable, but evolutionary theory is not. Why? The basic difficulty is that organic evolution, especially as it applies to man, touches very close to the individual's subjective sense of being and having a purpose which pervades almost all of our religious doctrines, formal and informal.

Evolutionary theory in its scientific setting is incompatible with this viewpoint. In its attachment to religious ideas, especially formal doctrines, the idea of purpose becomes associated with other concepts, in particular those of divine creation and an unchanging universe. So, although purposefulness, or *teleology*, is not in itself at all contrary to some sort of evolutionary theory, it may be associated with a broader philosophy which rejects the idea.

This discussion is much oversimplified. We might try to trace the ideas of teleology back to one of its early proponents, Aristotle, and apply it to his influence on the church. Or we might look at various

Figure 1 Four ancient Greek philosophers. The drawings of Plato and Aristotle from figures of busts in B. Russell (1959). Drawing of Heraclitus after figure in P. Tsankonas (1968).

rationalizations used to accommodate the sort of evolutionary changes we see in the framework of the formalistic scriptural world. Without an intensive investigation, however, the main point remains simple—there are two conflicting ideas, both of which have profoundly influenced the forming of our world view.

The idea of evolution often is close to the center of this conflict, touching, as it does on questions of just what man is, where he stands as an individual, what his role is and how he should behave in it, and what his purpose might be. We cannot investigate these questions without going into some reasonably difficult philosophical ideas. To appreciate how evolution relates to these ideas, we must have a good idea of what the general theory of evolution is and how the theory of organic evolution fits into it. That is what this book is all about.

Changing world — evolving world

Here we will consider the very broad idea of evolution. What we call *change* is a succession of events which results in different conditions with the passing of time. Evolutionary change is of this sort, and it carries three additional conceptual restrictions. One restriction is that the change is directional; it is the passage from one condition at one time to another condition at a later time. The second is that the change is gradual; there are no big jumps from one condition to another, but intermediate conditions always exist. The third is that the change is slow relative to our general sense of the passage of time. At one end of the time axis is change so slow that there appears to be no change, while near the other end is change so fast that it appears catastrophic. Evolutionary change lies between these extremes.

Successions of events, as we look at them in our time scale, seem to be of two sorts—those that are repetitive and those that are not. The first, in pure form, are cyclical. The rhythms of the body, succession of day and night, the succession of years, and the planetary motions are examples of cyclical changes. In each of these events, a regular return to initial conditions occurs. This change is not evolutionary. If everything, large and small and fast and slow, operated this way, there would be no evolution, for evolutionary processes do not involve return to an initial condition. Time in cyclical processes may be thought of as circular; in evolutionary processes, it is linear.

The matter now becomes more complicated. Cyclical changes obviously do occur, but if one wishes to be technical about return to precisely initial conditions, all sorts of bypaths, most of which we will avoid, open up. On a cosmic scale, as far as we can know from observations, there is a progression of events, a cosmic evolution. So while cyclical phenomena such as the motions of the planets in our own solar system seem to be repetitive, there is a slight amount of change over eons of time as the universe, the stars, and our own sun evolve. These events are not strictly cyclical, although for practical purposes within our time scale they are considered as such.

Up and down the time scale, sequences of events pass beyond our ability to comprehend them in the framework of common sense. Perhaps at a scale very much greater than we can recognize, all change is, in fact, cyclical. Perhaps at a very small scale, at the subatomic level, the basic entities are always the same, the only change being in their relative positions in a time-space coordinate system. In the time and size scale at which we live, however, the world does change. This change is linear and, therefore, evolutionary. Our changing world is fundamentally an evolving world.

Ways of looking at evolution

If the idea of slow, gradual, directional change is granted to be a fundamental property of the universe directly accessible to human observation, there still remains the question of how it operates or what causes it. If we start with the proposition that matter exists and that it acts under some general laws, it appears, as hinted before, that there are two ways of looking at what instigates change. One is teleological and incorporates the idea of purpose, the other is materialistic and holds that the physicochemical laws of matter are sufficient to explain all changes.

The *teleological* outlook sees the universe as the unfolding of some "grand plan". Change has a purpose—it is goal directed. In science, teleology is a "forbidden" word, but the idea of purpose certainly has appeal, since it fits into our conception of much that we see and do. It calls for an immaterial force—a spirit, a mind, a deity, a final cause—to give direction to what is happening. The motivation for change lies in the end result of all processes so that cause and effect, as we think of them in mechanical interpretations, are not strictly applicable. The immaterial motivator is acting continuously and was not just a force that created matter and sent it on its way. In biological evolution, a special life force may be called upon which may or may not be the same as a more general evolutionary force, depending on whether or not all matter is considered to have life qualities. We will examine this viewpoint, called *vitalism*, in more detail later.

The *materialistic* outlook is based on the premise that matter acts under the laws of physics and chemistry which are inherent in matter. The ultimate cosmic fact, according to this interpretation, is physical being, or material existence. No immaterial forces exist, and no purpose is considered. This is the materialistic interpretation.

The most common form of this idea has been a *mechanistic* materialism, under which the universe operates in a machinelike fashion. A cause precedes each event, and conditions at one time determine what the situation will be at a later time. This is a *deterministic* philosophy and is the fundamental philosophy upon which science was structured. Scientific thinking, however, has been restructured during the current century, so that laws of causality and determinacy have been replaced by the laws of *probability*; the scientist no longer pretends to be able to know the absolute truth. We will have more to say about this in the next section.

Like teleological doctrines, the mechanistic interpretation appeals to common sense, which is concerned with how things appear to operate. In biology, however, the denial of purpose is disturbing, for it, too, is part of our daily life. We will need to consider teleology several times later on but, for the most part, evolution as we will think of it will be cast in a nonteleological scientific framework.

A rather odd but interesting materialistic way of thinking about the universe is found in Marxian *dialectical* materialism, especially as it was applied to science by Engels. In this view, change is an integral part of reality. Change results from the conflicts of an event and its opposite, which exists as a necessary consequence of the event. Progress, or *synthesis*, is the natural outcome of the interaction of the two, but purpose is not involved and there is no goal. While it is materialistic, dialectical materialism is not mechanistic. Cause and effect do not follow the sequential patterns that they do in a machinelike universe. Followers of this philosophy (there are relatively few in science) feel that their universe includes an explanation for progressive change which cannot be found in a mechanistic universe. The concept of transformations in nature as a result of contraries interacting is a very ancient one. It is found, for example, in the Yin and Yang of Chinese philosophy, as expounded in 1200 B.C. in the *Canonical Book of Changes* (*The Yi-Ching*).

How one resolves these different points of view in his own thinking depends upon the basic assumptions that he wishes to make. The *pragmatist's* conclusion that the truth is best determined by the validity of the consequences of the concept offers one solution. This is a tenet basic to scientific philosophy and is a rationale of much that is said about evolution in this book. Whether or not a consequence is found to be satisfactory, of course, depends upon the intellectual climate in which the results are viewed. As the climate changes, so may the *truth* change. Ultimately, one must make his own decision as to what fulfills his intellectual and emotional needs most adequately.

Biological evolution

The most general theory of organic, or *biological*, evolution holds that all forms of life differentiated from an initial life stock by gradual changes. All beings, from bacteria to men, trace back through unbroken lineages to the beginning of life. Tens of millions of species have come into existence and subsequently perished. This is the course of evolutionary progress and, without question, all the contemporary species will also become extinct, although in future times their altered descendants may thrive.

Evolutionary theory explains, or integrates, a vast amount of information. It is the silver thread of present-day biology, incorporating data from *morphology*—the science of form, *taxonomy*—the association of organisms into meaningful groups, from *developmental biology, genetics, biochemistry, physiology, ecology,* and so on. The fossil record of extinct species indicating their arrangement in time, progressing from simpler to more complex forms, adds an important dimension. A true sense of the duration of time comes from this record as well.

Successive populations of plants and animals, as viewed today, undergo progressive changes, some of which are of an evolutionary sort. Such changes occur without man's intervention in nature, but may be hastened by man in the breeding of domestic animals and in experiments under controlled laboratory conditions. Evolution by small changes is shown to occur under these conditions; it is an observable fact. That this sort of evolution took place throughout the full history of life is inferential, but the inference is strong. Was gradual evolution the only means of change? As far as we can tell, it was, since no other kind of change has been found among living organisms, and since all the evolutionary phenomena we know *can* be explained in this way. It does not necessarily follow that there were no other kinds of change.

Acceptance of the general theory of organic evolution does not involve adherence to any particular philosophy of the mechanism or motivation of change. Both vitalists and mechanists can find explanations that satisfy them. Evolution can be fitted into all but the strictest interpretations of the Bible. When we try to go beyond description toward explanation, however, limits begin to develop, but no theory receives very wide acceptance unless this step is taken.

Darwin's idea of evolution by means of *natural selection* provided an explanation. The basic concept behind natural selection is simple, and can be reduced to three main components. First, the individual members of each species differ from one another; that is to say, there is variation within species populations. Many of the differences are heritable and passed on from parents to offspring. Second, some of the varieties are more successful than others. These are said to be better adapted to their environment. Third, the more successful individuals produce more offspring, this in itself being a measure of their success. The next generation, by this same process of natural selection, includes a higher percentage of the successful variants and is better adapted to the environment. The fact that many more young are produced in each generation than can possibly survive to reproduce adds vigor to the selective process. There are many complications, but we will get to these later in the book.

Natural selection is a mechanistic explanation of organic evolution. This brought it into conflict with the prevailing religious temper of the middle of the nineteenth century, almost immediately upon publication of Darwin's book, *On the origin of species by means of natural selection*, in 1859. Controversies naturally made the theory better known and were partly responsible for its remarkable social impact.

Darwin's arguments came from everyday common-sense observations, expertly marshalled and presented. These could easily be understood by layman and scientist alike. Darwin's mechanisms of change involved no nonmaterial force, and his conclusions could be tested against actual circumstances. In the nineteenth century, society was ripe for this sort of a theory, one which fitted the mood of inevitable progress that was abroad throughout Europe and North America. In spite of widespread acceptance, however, dissent has continued to the present time. Arguments on vitalism versus materialism lie at the heart of the matter.

To see organic evolution in its scientific setting, we need to delve a little deeper into some of the matters touched upon earlier with regard to evolution as a whole. When Darwin's theory was introduced, science was experiencing a period in which mechanistic and deterministic philosophy was in full sway. Observations of the order of universe were *known* to produce meaningful answers. The first cracks in the scientific monolith were only beginning to form; as research has continued to the present day, some very basic concepts

BIOLOGICAL EVOLUTION

have been weakened. The nature of matter has become somewhat of an open question to be treated differently at different levels of organization. Determinism has been modified, prediction now being only statistical in many cases. An element of indeterminacy has been found at subatomic levels.

Criticism of scientific materialism has also come from sources outside science, and this sort of criticism long preceded Darwin's time. It is argued, properly enough, that what we recognize by our senses as the *world out there* is a severe abstraction from the whole. Our concept of the world may not be wrong, but at best it's only partial. Further, our sensory impressions are far removed from the objects sensed. Our mental impressions when we see, hear, smell or feel something are the result of a long series of complex energy exchanges which lie between the sensory receptors and the conscious image. Many philosophers have concluded that a straightforward materialistic view of the world is incomplete at best and totally false at worst. If these critics are right, where do matters now stand: Should we simply give up? Of course not! Every way of gaining knowledge, including the *rationalistic* approach which leads to questioning the reality of the material world, makes some assumptions and incurs some restrictions in making them.

As far as organic evolution is concerned, we can look at it like this. Man views his world in his own particular way. He has become adapted to this world and to his particular way of viewing it and reacting to it by natural selection. His view of the world may differ sharply from that of other organisms, but it best depicts the one necessary for his survival.

As technology has advanced, the world to be understood by man has expanded. With it, philosophy has broadened. Some of the simpler materialistic mechanistic truths have been shaken. So far, the theory of organic evolution has been little affected. The complexity of living matter is so great that atomic and subatomic phenomena, where problems such as indeterminacy or relativity arise, have had no great impact. The rising interest in molecular biology may one day alter this. The basic changes in philosophy alone inevitably have modified, to some extent, the way biological processes are viewed.

Even so, it is at the cellular and supracellular levels that evolution by natural selection operates. At these levels, largely comprehensible under the view of the world to which man became adapted long ago, the mechanistic theory of natural selection can explain all evolutionary phenomena adequately. Changes may come about in the future. The direct impact, it would appear, will first occur at subcellular levels where the variations that supply the grist for the mill of natural selection originate.

IMPORTANT CONCEPTS

CHANGE, CYCLICAL (*or* CIRCULAR): change involving repeated return to initial conditions. Describes planetary motions, body rhythms, etc.

CHANGE, PROGRESSIVE (*or* LINEAR): change involving continuous development toward a different (and not necessarily higher or *more perfect*) state of being. Evolutionary change is progressive.

COMPLEXITY: the state in which distinguishable parts are related in such a way as to give unity to the whole. The more numerous the interrelated parts, the more complex the system.

CREATION: that doctrine which ascribes the origin of matter and distinct species of organisms to acts of creation by God.

DETERMINISM: the doctrine that every process in the universe is guided entirely by natural law. All events are absolutely dependent upon and conditioned by their causes.

EVOLUTIONISM: the doctrine that the universe, including inorganic and organic matter in all its manifestations, is the product of gradual and progressive development.

EVOLUTION, BIOLOGICAL: the process of gradual and progressive development of organisms resulting in diversity of species.

INDETERMINACY: the state of having inexact limits, or the condition in which complete quantitative measurement of dimensions, states, or processes is impossible.

MATERIALISM: the doctrine that everything is explainable in terms of matter and energy alone, and that all qualitative differences are reducible to quantitative differences.

MATERIALISM, DIALECTICAL: the doctrine that change occurs as the result of the conflict of opposites; that is, that the conflict of thesis and antithesis produces a synthesis. As in strict materialism, all changes have their origins in the properties of matter and energy.

MECHANISM: the doctrine that all phenomena, including organisms, are totally explainable by mechanical principles. Nature, like a machine, is an entity whose single function is served automatically by its component parts.

NATURAL SELECTION: the differential action of environmental factors upon the variability within a population or species tending to eliminate those individuals which are less fit.

PHILOSOPHY: a study of the processes governing thought and conduct. The theory or investigation of the principles or laws that regulate the universe and underlie all knowledge and reality.

IMPORTANT CONCEPTS 11

PRAGMATISM: the doctrine that finds the meaning of a proposition in its physical or logical consequences rather than in the proposition itself.

PROBABILITY: the application of mathematical methods to the determination of the likelihood of an event when the data are not sufficient to determine its occurrence or nonoccurrence with certainty.

RATIONALISM: a doctrine in which the criterion of truth lies not in the sensory perception of material objects and processes, but in intellectual and deductive processes.

SCIENCE: systematized knowledge derived from observation of, study of, and experimentation on physical processes and material objects.

TELEOLOGY: a doctrine that explains the past and present in terms of the future. All events occur toward an ultimate purpose, goal, or final cause.

VARIATION: within a species, the deviation of offspring from the parental type and from each other in any biological characteristic. Variability has its origin in the genetic materials.

VITALISM: doctrine that organic beings possess a vital force, or *elan vital*, that is not explainable in terms of physical and chemical laws.

REFERENCES

Appleman, Phillip. 1970. *Darwin*. New York: W.W. Norton & Co., Inc.

Barzun, Jaques. 1958. *Darwin, Marx, Wagner: Critique of a heritage.* New York: Doubleday & Co., Inc.

Dewey, John. 1910. *The influence of Darwin on philosophy.* New York: Henry Holt & Co.

Eiseley, Loren. 1961. *Darwin's century.* New York: Doubleday & Co., Inc.

Greene, John C. 1959. *The death of Adam.* Ames, Iowa: Iowa State University Press.

Himmelfarb, Gertrude. *Darwin and the Darwinian revolution.* New York: W.W. Norton & Co., Inc.

Mason, Stephen F. 1962. *A history of the sciences.* New York: Collier Publications.

Runes, Dagobert D. 1967. *Dictionary of philosophy.* Totowa, New Jersey: Littlefield, Adams & Co.

Russell, B. 1959. *Wisdom of the west.* New York: Crescent Books. pp. 59, 80, 270.

Schneer, Cecil J. 1969. *Mind and matter: Man's changing concepts of the material world.* New York: Grove Press, Inc.

Tsakonas, P., ed. 1968. The origin of medicine in Greece. *The European Congress of Cardiology.* Athens. p. 26.

Wilshire, Bruce. 1968. *Romanticism and evolution: The nineteenth century.* New York: G.P. Putnam's Sons, Capricorn Books.

2 | The nature of life

Perspectives

Everyone has his own subjective view of what life is. Some humanists insist that this personal observation can be the only rational source of information about life, since the analysis of scientists involve abstraction and reduction and are thus inevitably out of touch with life's *essence*. Strictly speaking, of course, such a subjective view relates only to human life, opinions concerning other life forms being inferential. Because each person's view of life in general is colored by his thoughts and feelings about his own life, the search for an understanding of *life* as a whole has led to conclusions almost as numerous as serious students of the matter. Even today, nothing is very well settled in this area, but the different views are important to us because they have profoundly influenced the ways in which the various aspects of evolution have been and are regarded.

If we speak in more objective terms, distinctions between living and nonliving matter are usually reasonably clear. The life substances, processes, and organizational patterns seem to be very different from those of nonliving matter. However, there are borderline cases. When living matter dies, there is the problem of determining

ARTICULATORS OF THE THEORY

ALFRED WALLACE
Advocated evolution as the mechanism of biological change at the same time as Darwin, but on mostly theoretical grounds.

CHARLES DARWIN
Amassed an impressive amount of evidence for natural selection and the evolutionary origin of biological species.

EXTRAPOLATORS OF THE THEORY

FRIEDRICH ENGELS
Merged evolutionary thought with dialectics and applied these principles of social and political change.

PIERRE TEILHARD DE CHARDIN
Merged evolutionary thought with religious ideas about the perfect end point of biological and spiritual change.

the exact time of death. Viruses show only some life properties, and these only part of the time. Crystals can grow and exhibit organizational properties somewhat like those in living systems. These instances, emphasizing the indistinctness of the boundary between the living and the nonliving, underline the fact that life is recognized by its processes and activities, as well as by the nature of its constituent matter. They also hint at the difficulties that accompany the widely held notion that matter and living matter are somehow distinct. Whether this distinction is real or not, it has played an important role in science. It is the reason that we have two major disciplines in natural science, the physical and the biological. The distinction is also important in our everyday thinking, for it is an intuitive notion which is hard to disregard.

If we hold to an evolutionary theory of the universe, it becomes apparent that there was a time during the earth's history when life did not exist and then a period during which it developed. The earliest known fossils are between 3.0 and 3.5 billion years old, so that the origin of life and the beginning of organic evolution date at least this far back. The details of the transition and the relationships between living and nonliving matter thereafter are critical to an understanding of the evolutionary process and will be more extensively dealt with in a later chapter. For the present, we will limit ourselves to the examination of various philosophical and scientific viewpoints dealing with the nature of life itself, with a critical eye to the biases each introduces into an evolutionary frame of reference.

Scientific points of view

When the life-matter problem is placed in a materialistic scientific-context, it begins to have well-defined limits. Life exists only in association with matter; it has a material base and can be recognized only as a manifestation of matter. Activities on the organizational levels at which matter shows life properties are deterministic and machinelike. With such limits imposed, it becomes possible to come

Figure 2 Founders and extrapolators of the theory of evolution by natural selection. Drawing of Darwin at age 40 based on photograph of painting from facsimile of first edition of *On the origin of species*. Drawing of Teilhard de Chardin based on photograph in J. Hemleben (1966). Drawing of Engels after photograph in B. Russell (1959). Drawing of Wallace after photograph taken in 1878, in A.R. Wallace (1905).

to grips with problems of the nature of life and to arrive at testable hypotheses concerning its properties.

During the last few decades, through studies of the physicochemical properties of organic and living matter, scientists have made great strides toward determining the physical bases of some of the most important attributes of living matter. In particular, *metabolism*—the chemical reactions carried on by living organisms—and *replication* have come to be better understood. Knowledge is still partial, but the way has been opened to a much more complete insight of the nature of life's processes than was ever imagined to be possible.

Most of these advances have resulted from the standard sort of scientific analyses, taking advantage of new instruments and techniques. Structures have been isolated and studied as simple entities with powerful light microscopes, electron microscopes, X-ray machines, and similar instruments. Radioactive minerals, introduced into systems as tracers, have been used extensively to elucidate biological processes. Faith in the results of these analyses is based on the proposition that the phenomena of life, known at macroscopic levels, can best be understood by study of processes at microscopic and submicroscopic levels. This approach, used also by physicists in their investigations of the nature of matter, is called *reductionism*. Because significant events are studied alone, a large amount of integration and synthesis is necessary for understanding more complex phenomena.

What is emerging from the reductionist approach is a very precise way of viewing life as a spectrum of functional properties dependent upon a particular common material base for their existence. The base is formed by relatively few essential chemical elements among which carbon, oxygen, nitrogen, hydrogen, and, to a lesser extent, phosphorus and sulfur, predominate. A few classes of large, complex molecules formed from these elements, for example, proteins, nucleic acids, and carbohydrates, are the fundamental materials of life. The chemical activities of these molecules are, or produce, the properties of living matter. Inasmuch as the basic properties of life can be explained by these physical and chemical laws, nothing further is necessary.

Living matter, biochemically speaking, is any matter that both metabolizes and reproduces itself. The other attributes we may associate with life—movement, sensitivity, communication, and so on—are either manifestations of the two basic attributes or follow directly from them. This provides a satisfactorily concrete and operational concept of matter and life which many scientists find fully adequate.

SCIENTIFIC POINTS OF VIEW

Several problems do arise from such a definition of living matter, however. One is whether a definition based on the *attributes* of life is sufficient to differentiate living from nonliving matter. Can we not find *inorganic* systems in which similar processes occur? Phenomena very similar to growth, utilization of environmental resources, and replication can be seen in nonliving crystals, especially in aperiodic (asymmetrical) crystals. Were we to construct a machine that performed all the basic life functions, could it be considered alive? It is obvious that living and nonliving are not clearly distinct, especially since a fair case can be made for the proposition that the processes we associate with living matter are not unique to systems we recognize as alive.

A second problem which tends to arise from a strictly biochemical view of life is whether by defining matter or life on the basis of special properties, we are, in fact, defining what we mean when we speak of life. The reductionist approach states the equation, *life equals properties of living matter*. This approach leaves no place for the subjective concept of a unified sense of life. In view of such questions, is it really appropriate to use the word *life* in a scientific context?

Perhaps it is not valid to reduce all the attributes we subjectively think of as pertaining to life to a least common denominator including only those properties common to *all* life, as we do when we claim to understand life solely in terms of its physicochemical bases. This does not mean that other attributes of life contradict the physicochemical explanation or cannot be derived from a common base; it does raise the question of whether, in the course of abstracting, we may not be losing touch with some critical aspects of life, as the humanists suggested. These and other related questions arise continually in biology, and students in different areas of the discipline have arrived at substantially different answers to them.

Scientific evolutionary studies analyze the physical changes of organisms and the changes in their activities, and operate under the assumption that all form and function can be understood by viewing life materialistically. Strictly mechanistic interpretations, which are strongly reductionist in character, find a much more mixed reaction. Current investigations of the genetic basis of behavior in individuals and populations are, however, reinforcing the feeling that even the most intangible aspects of organisms and their relationships will eventually yield to studies founded in determinism.

Still, much remains unknown. This lack of knowledge seems to result more from the complex nature of the investigations and from failure to ask the right questions than it does from adhering to inadequate basic concepts. The problem is particularly acute in the case of intangible phenomena which are not fully defined or which seem to lack a materialistic base, such as extrasensory perception (ESP) or problems related to the existence and nature of the mind. The scientific attitude toward such phenomena is mixed. At one end of the spectrum is outright rejection of their existence; at the other is acceptance, but with reservations, pending a deterministic explanation. Most biologists prefer to adopt a wait-and-see attitude, reserving opinion until further evidence has been presented.

The phenomenon of life falls into this gray area. A clear scientific definition that also satisfies the intuitive sense of life seems, at present, to be impossible. In scientific studies, the concept of life may be just a convenient, temporarily useful, but ultimately fictional, collective. It may merely stand for a spectrum of activities that is a normal consequence of a particular organization of matter acting in accord with the known laws of physics and chemistry. Perhaps, rather than thinking of the evolution of life, we should think only in terms of living *matter*.

Throughout all but the final chapters of this book, we will take up evolution from a scientific point of view. The scientific approach has been a fruitful one, and only after it developed was evolution subjected to serious systematic study. It is important to keep in mind, however, some of the many philosophical views of life and matter, since our scientific thinking is unavoidably influenced by them. The rest of this chapter is devoted to a brief survey of these concepts.

Philosophical points of view

We have already dealt briefly with the contrasting views of vitalism and materialism and have made the point that, although science is founded upon the latter, the vitalistic concept has considerable popularity. Here vitalism means that life is a specific, nonmechanical, and nonmaterial entity, and this definition serves to distinguish it in an unmistakable way from materialism.

In everyday usage, the line between vitalism and materialism is much less distinct. Even some materialistic scientists have favored a vitalistic concept of life, calling life a property separate from matter. Claude Bernard, as stated in his *Definition de la vie* (1878), for example, felt that neither materialism nor vitalism alone could give a satisfactory explanation of life. He reasoned that the molecular changes characteristic of living systems, although physical, take place only in living systems—hence, they are vital. Although they are different from nonvital processes in nonliving systems, they do not result from nonmaterial influences. Bernard's concept has been called *physical-vitalism.* Is it vitalistic or not?

Any interpretation in which life is treated as something apart from or more than strictly inanimate physical activity has the chance of being called vitalistic. In the extreme, any concept which is not reductionist may take on a vitalistic aura. Within the past few years, a new point of view, called *holism,* has become more and more popular. The scientists who subscribe to holism are materialists, but they feel that organisms, populations, and the like should be studied as entities, since the laws that apply to higher levels of organization are not totally physicochemical in nature. The holists prefer to apply systems theory and information theory, *cybernetics,* to the study of life. Trouble arises because, in such cases, although it is not always justified, the implication of mysticism may creep in. The term *vitalist* as applied to a scientist by his colleagues usually carries derogatory overtones.

A useful way of looking at the many different interpretations of life and living matter and how they affect evolutionary thinking is in the perspective of the *relationships* of life and matter.[1] Most of the main ideas can be grouped under four general headings.

Emergence

Life is an *emergent* consequence of the *organization* of matter. Such attributes as motion, reproduction, metabolism, and sensitivity represent an ensemble of activities characteristically performed and which may be collectively thought of as life. But these activities are *not* predictable on the basis of physicochemical structures and laws, so that life, because of its organization, is more than the sum of its component parts. Emergence can easily be a materialistic concept and not out of keeping with scientific materialism, but it is not mechanistic.

The reductionist interpretation of life does not fall within this category. The majority of whole-animal biologists, when they think about it at all, tend to follow an emergent line of thought consistent with their holistic approach to science. Molecular biologists, however, rarely think this way. The principal factor influencing the holist is that the phenomena he studies involve complex levels of organization

---

[1] Thomas S. Hall has given a detailed analysis of the history of views on life and matter in his two-volume work, *Ideas of life and matter.*

which are not directly accessible by way of physics and chemistry; even though reduction to physicochemical laws might be possible through a series of steps, the result would be of no immediate value.

This may be applied, for example, to studies of complex goal-directed behavior such as the search for food by an active predator. To attempt to reduce his motivations, his actions, and his satisfactions to energy exchanges, molecular activity, or to the second law of thermodynamics is to lose the perspective of his role as a functioning organism and his role in the community as a whole. Total energy flow through the ecosystem of which he is a part might be so treated, but not the uniquely living interactions of the predator himself.

The concept of life as considered in the light of dialectical materialism is distinctly emergent. Life is looked upon as an inevitable unfolding of *the full tree of life* from beginning to present, and as such cannot be defined by any one stage in the series. The spiral of progress insures that life has evolved and will continue to unfold new forms, since that is the nature of change. Matter, since it is continually in motion, passes through progressive evolutionary stages to new and more complex forms of motion. As a result, new properties of matter arise. Life is a special form of motion of matter; the laws of chemistry and physics applicable to prebiological matter do not apply, although they are not violated. With life, something new has emerged.

Immanence

Life is *immanent* with respect to matter. It is either identical to matter or concomitant with it, as is, for example, gravity. If this position is taken, then all evolution—inorganic and organic—is a manifestation of the same general process. Nothing new emerges because everything existed actually or potentially at the start.

Many prominent scholars of the past have held this general idea. Spinoza of the seventeenth century and Robinet of the eighteenth both considered matter animate. Denis Diderot, a contemporary of Spinoza, was an evolutionist whose evolutionary ideas and concepts of life were closely related. He saw all matter as *sentient*, or conscious, some inertly and some actively so. No hard-and-fast line separated the living and nonliving, but the sentient aspects became increasingly evident in the more advanced forms of matter. Evolution, then, involved an increase in the degree of expression of this particular quality.

PHILOSOPHICAL POINTS OF VIEW

The idea of immanence avoids some of the mystical connotations of emergence, and this characteristic has made it attractive. Some rather difficult philosophies have developed around the sense of inseparability of life and matter. Henri Bergson, especially in his *Creative evolution*, and Pierre Teilhard de Chardin in *The phenomenon of man* expressed evolutionary views which followed this line of thinking. Both have been harshly treated by evolutionary scientists, but they have made contributions to the ideas of life, matter, and evolution that have had widespread effects in nonscientific areas.

Bergson's evolutionary thinking, as succinctly described by C.E.M. Joad in his *Guide to philosophy*, represents an alternative interpretation of the empirical base of scientific materialism. He began with materialistic data, but found mechanistic explanations of life and evolution inadequate. The universe, as he viewed it, is a continuous flow originating from a constant source; evolution is simply the movement of this flow. The universe is the creation and expression of a vital force or impulse whose function is to continually change and evolve. The essence of Bergson's philosophy is his concept of duration, which is his vital spirit, the *elan vital*.

This, of course, is a vitalistic philosophy, and it is difficult to relate Bergson's ideas to those based on materialistic premises. He sees matter as introduced by the intellect which "stops" the flow of reality and creates a common-sense notion of solid things. Further, matter appears as a counteraction, a back-flow of the normal flow of reality. This he likened to drops in the descending waters of a fountain (in his much-quoted analogy, *life is like a fountain*) with its continual upsurging spray from the eternal source and the downward countercurrent which gives the impression of stability in the upward-moving system.

Bergson's is a totally evolutionary philosophy which enjoyed immense popularity in its day, but it has now been reduced to obscurity. The main objection is to the contention that matter is merely an illusory part of the whole made by intellectual crosscuts of flowing reality. This philosophy seems fanciful to an objective materialist, for much of it is foreign to what common sense dictates.

At about the same time Bergson's ideas were having their impact, however, scientific study was beginning to revolutionize man's outlook on the universe. There emerged the *relativistic* world of Einstein in which distance, duration, and derived qualities are relative, dimensions change with velocity, space is curved, and gravity is a field phenomenon. The relativistic world is no more self-evident to one schooled in the context of an Euclidean universe than is the universe of Bergson. Einstein's view is empirically based and subject to experimental verification, but it extends credulity nonetheless. The concept of an expanding universe with continuous creation of matter, as postulated by the eminent cosmologist Hoyle and others, is strongly reminiscent of Bergson's fountain analogy. It appears that *truth* need not be self-evident, and that theories need not be rejected simply because they extend beyond the boundaries of common sense.

Teilhard de Chardin was a paleontologist and a Jesuit whose best-known work, *The phenomenon of man,* was published in the 1950s and has caused much scientific, as well as philosophical, controversy. Unfortunately, it has many internal inconsistencies, so that it is difficult to know just how his ideas of life and matter are to be understood. The chapter of his book entitled "The advent of life" seems to recognize life as an entity with emergent properties. Life began, he states, with the cell. However, he also finds that nonliving molecules are incomprehensible unless they are assumed to contain a rudimentary psyche. In this sense, life seems immanent. The coming of life is an unfolding of what preceded it—an elaboration of immanent properties—but at the same time, it is something more. Evolution, as he viewed it, is a characteristic of all matter; it passes from lithosphere to hydrosphere to biosphere, and finally to the noosphere, the sphere of intellect. His work includes an odd mixture of emergence and immanence, physical science and mysticism. Teilhard's philosophy has had considerable impact in the last twenty years and illustrates the effects of groping for truth under the dual influences of materialistic and vitalistic concepts. Regrettably, the clarity of neither the vitalist nor the materialist is to be found in the final product.

Even more remote and difficult to comprehend is the philosophy of Alfred North Whitehead, one of the most prominent philosophers of the present century. A few comments on his position are included to show the direction that a philosophically-oriented scholar may take when he thinks beyond the confines of empirical science. The whole universe in Whitehead's cosmology is living and creative, and all entities are related and intercommunicative. The concept of universe-as-organism goes very far back in philosophical lore to the Greeks and beyond, but Whitehead's organismic cosmos was at a different conceptual level. He considered the division of life and matter to be one of the most misleading and disastrous of dichotomies, for life and matter are the interwoven threads of that which is the

PHILOSOPHICAL POINTS OF VIEW

universe. This is an example of holism and immanence on a cosmic scale.

Whitehead's ideas also contain an element of emergence, however. Life represents the origination of a novelty of character; an organism is alive when its behavior cannot be explained entirely in terms of antecedent events. Life to him also has an element of mentality and is intimately interwoven with his concepts of society. Society, though, does not carry the usual meaning, for it is part of the whole scheme of the universe with all of its parts intersentient.

The foregoing concepts are all evolutionary, but they have little to do with common sense and mechanistic understandings of the subject. As the old established scientific philosophies have become less satisfactory in the light of our new knowledge of the universe, the search has begun for new general concepts, with Whitehead and Teilhard being only two men who have attempted the difficult union of science and philosophy.

Life as immaterial force

Life is an immaterial force which is *imposed* on matter. The viewpoint that external forces influence the behavior of matter has been important in the past and is widely held today. It forms the basis of many forms of religion, especially in the recognition of life—spirit or soul—as an attribute special to man. Applied to life, the concept is strictly vitalistic.

There are two forms this concept commonly takes. One is that life essence exists only in association with matter, bonded, so to speak, with substance. The other holds that the life essence can exist independently of matter. The first approach has some elements of Bergson's *elan vital,* but it differs in the recognition of the life force and matter as distinct and of matter existing totally devoid of life or psyche.

The existence of an immaterial life force is, of course, inadmissible in materialistic science. If we do accept the duality of material and immaterial energy, however, and incorporate the idea of an imposed life force, we must then modify the scientific conception of evolution. Nonliving matter cannot be motivated by this second force. Somewhere there must be a line between the materialistic and nonmaterialistic regimes. At just what stage did this distinction occur? Was it when the first cells arose? Was it only at some more complex level? Can matter evolve only under the influence of the life force?

Did the force overcome the limitations of organization that otherwise restrict all matter to mere inanimate activities? If this were the case, no scientific study of evolution could be possible. If, however, physical laws still hold in the presence of this force, then physical analysis of physical change remains possible. This approach always finds itself confronted with a paradox; all we can know of material changes comes from observations, and we base our thinking on the premise that this is insufficient.

As far as the immaterial force, or *essence*, is concerned, all sorts of evolutionary questions may be posed. Is there one essence or many? If there is only one, as is usually supposed, does it evolve? Is it a higher form in more complex organisms, in mobile beings as opposed to more sessile ones, or in intelligent or conscious ones? Is such existence a primitive state, and does it evolve to become free of matter? These questions cannot be answered within the limits of materialistic science and are not testable by empirical means. They may be both rational and logical, however, starting from the *a priori* of the existence of an essence.

Life as organization

Still another way that life has been viewed is that it is *organization*, such that matter behaves in a lively fashion. The lively behavior is not life; rather, life *is* organization. This idea, too, has its roots deep in the history of man's concern with the organic world. It is a fundamental point in Aristotle's considerations of matter and form. The distinguishing structure of a thing is its form, and it is in accordance to that form that matter is organized. Living matter is that whose organization can extend and propagate itself by imposing its organization on other suitable matter; this it does through growth and replication. Thus, living matter not only *is* organization, it *creates* organization.

This general concept is one found in partial form in various other considerations of life and matter. Lamarck (1744–1829), for example, held this view, but he also added an emergent aspect. He effectively bridged the separate ideas of life as organization and life as a physical manifestation of organization. There are few proponents of this idea today. Although organization is considered to be a major contributor to the phenomenon of life, it is not thought to be life itself. Organization rather *leads* to life or is accompanied by life processes such as motion, metabolism, and replication.

From the foregoing discussion, two things are clear. One is that man's search for understanding of the universe has included a search for the nature of life, and these two cannot be separated. Secondly, this search has not resulted in a definition of life that is generally satisfactory. Life may be defined in numerous ways for many purposes, and it is by no means true that life in different contexts is the same thing. The word is a useful collective, but search for a common understanding is futile, because the concepts grouped under the collective do not all proceed from the same basic premises.

Perhaps one of the best ways to view the life-nonlife dichotomy, if it really exists, is in the context of the actual origin of life. By looking at this series of events and the evolution that followed, and by asking the proper questions, we can perhaps clarify some of the mystery, philosophical and otherwise, which surrounds our concepts of life.

IMPORTANT CONCEPTS

ANALYSIS: the separation of anything into its constituent parts, or the examination of anything to distinguish its parts separately or in relation to the whole.

A POSTERIORI: arriving at principles by generalization from facts, or *after the facts.*

A PRIORI: deducing consequences from self-evident definitions or principles. Designates that which can be known by reason alone.

COSMOLOGY: a point of view in which the universe is treated as a harmonious, orderly system which is governed by equally harmonious laws. (This is opposed to the view that the universe is chaotic.)

EMERGENCE: the doctrine that *life* is more than the sun of its physicochemical components and processes; it is the result of organization of matter and partitioning of energy in a characteristic manner.

EMPIRICISM: the pursuit of knowledge by observation and experiment. The philosophical doctrine which attributes the origin of all knowledge to direct experience.

HOLISM: a point of view in which systems are to be studied as such; that is, there are laws which are applicable at levels above the physicochemical, applying to states of organization. (Opposed to reductionism.)

HUMANISM: any view in which the welfare of mankind is of primary importance. The human mind is often taken to be the source of all truth, since all known interpretations of truth are undertaken by humans.

IMMANENCE: the doctrine that *life* is an inherent property residing in all matter, but is expressed more or less strongly at different levels of organization.

OBJECTIVE: dealing with facts independently of mind and human interpretation.

REDUCTIONISM: a point of view in which analysis of all the physicochemical components and processes in a system is sufficient to explain all the properties of that system. (The biochemist's viewpoint.)

RELATIVISM: the idea that there is no absolute standard against which all other phenomena can be compared. To compare phenomena, a frame of reference must be constructed, and this introduces a bias in the outcome of the investigation.

SENTIENCE: the property assigned to a being capable of sensing and consciousness.

SUBJECTIVE: dealing with facts arrived at by mental processes and interpretation, based ultimately on the point of view of the individual.

IMPORTANT CONCEPTS

SYNTHESIS: the combination of separate parts or facts to produce a more complex entity or concept.

THERMODYNAMICS: science which deals with the properties and interactions of energy. It is based on two laws:
1. The amount of energy in the universe is constant.
2. The *entropy* (degree of randomness) in the universe is increasing.

REFERENCES

Darwin, C. 1967. *On the origin of species.* A facsimile of the first edition. New York: Atheneum. Frontispiece.

Hall, Thomas S. 1969. *Ideas of life and matter.* 2 vols. Chicago and London: University of Chicago Press. pp. 399, 419.

Hembleben, J. 1966. *Selbstzeugnissen und Bilddokumenten.* Hamburg: Rowohlt. p. 6.

Joad, C.E.M. 1936. *Guide to philosophy.* New York: Dover Publications. p. 592.

Mays, W. 1962. *The philosophy of Whitehead.* New York: P.F. Collier, Inc. p. 288.

Teilhard De Chardin, P. 1961. *Phenomenon of man.* New York: Harper & Row, A Torchbook. p. 318.

Wallace, A. R. 1905. *My life. A record of events and opinions.* vol. 2. London: Chapman and Hall Ltd. p. 98.

Whitehead, A.N. 1925. *Science and the modern world.* New York: The Macmillan Co. p. 191.

3

The origin of life

Perspectives

If it were possible to construct a cell from scratch using only inorganic materials and supplying it with energy sources of the appropriate variety, would it be truly *alive*? The answer is almost unequivocally yes, unless we conclude that some supernatural force motivates life. It is a somewhat frightening answer, or at least an awe-inspiring one. In one way, however, it gets right to the heart of the problem of what life really is; and it doesn't really matter whether life was created in our test tube or in nature's "laboratory" many aeons ago. So far, we have fallen far short of the goal of synthesizing a living system; but with her almost unlimited time, nature seems to have turned the trick.

Once living matter appeared on earth, the principles of Neodarwinian selection began to shape the processes of evolutionary change. We know a great deal about this evolutionary change. However, its forerunner, *prebiological* evolution, involved molecular selection based on the properties inherent in the prebiotic organic molecules themselves; the agent of selection was chemical stability and the establishment of localized chemical equilibria. These events occurred remotely in time, more than 3.2 billion years ago, and left almost no interpretable record. Some clues about the nature of ancient environments are at hand, but we are forced to rely upon living organisms and experimentation to plot the ways in which the events leading to the origin of life may have gone. At the present time, there are several plausible models explaining the circumstances under which life came to be, but as yet there are no reliable criteria for choosing the best alternative. Ideas and models concerning the origin of life continue to evolve as new experiments are performed, new questions are asked and more information from ancient rocks is discovered. Our approach, then, can be twofold. We can examine chemical evolution from the vantage point of modern biochemistry and cell biology and also from the records written in the rocks by the earliest known living forms.

The unity of structure and function

The *cell* is the basic unit of all living matter. Many structures, substances, and mechanisms are the same or closely similar in all kinds of cells. Even the two fundamental kinds of cells, *procaryotic* and *eucaryotic*, are alike in minimal composition and in the carrying out of vital functions, even though their superficial appearances are very different.

In all cells, for example, information transfer in replication is by nucleic acids, and catalysis, necessary for metabolism at low temperatures, is carried out by certain proteins, the enzymes. Smaller molecules involved with energy-exchange cycles are similar throughout the living world. Further, only twenty different amino acids occur in cells, although there are many others to be found in nonliving matter. Those in cells, with rare exceptions, are all *L*-amino acids; that is, their structure is such that they polarize radiation in a "left-handed" fashion. Inorganically prepared mixtures of common amino acids, on the other hand, usually include about equal proportions of right-handed and left-handed molecules.

Although the basic chemical processes are uniform, the variety of form, or *morphology*, of the organisms carrying them is immensely varied. Morphological differences are, of course, accompanied by functional differences. Thus, at the molecular level, the *monomers* and energy-exchange molecules are the same, but differences arise in the arrangement of molecules in *polymers*. The result is that the *macromolecules* exhibit great diversity, even though the kinds of their constituents is strictly limited. This is true, for example, for proteins, which are formed by polymerization of amino acids and which contain these characteristic molecules in different sequences and proportions. The enzymatic actions of the proteins differ, and this is a direct result of structural differences resulting from amino acid sequence differences. Much the same principle applies to the formation of nucleic acids from nucleotide sequences; changes in coding and transmission of information occur when the arrangement of the constituent molecules is altered.

The common structures and processes, along with the diversity of higher-level structures derived from them, came into existence by prebiological (or precellular) evolution. The important events in this process probably included (1) the evolution of chemical processes characteristic of cellular systems, (2) the selective preservation of reaction products at various levels, and (3) the association and concentration of these products into discrete self-replicating cells.

Prebiotic evolution must have been a time of extensive chemical experimentation with many different kinds of systems arising and perishing under the eye of chemical selection. By the time that cells had come into existence, however, the metabolic and replicative patterns had been pretty much decided upon, and the unity of structure and function of these earliest cells has been passed down essentially unchanged to modern forms. *Cellular* evolution, then, has proceeded in three phases: (1) initially great chemical variety in the form of small organic molecules; (2) reduction of variety by molecular selection, chemical equilibria favoring certain reaction products over others; and (3) development of a uniform biochemistry with the development of living cells. This last phase provided the basis for Neodarwinian natural selection acting upon the organized system as a whole, since it is the presence of relatively uniform genetic materials that makes Neodarwinian selection possible. As errors in the genetic information accumulated, new *phenotypes* arose and led to the great flowering of structural and functional diversity that followed. In the rest of this chapter, we will take up the first phases of the sequence—those leading to the formation of the cell. Later chapters will treat the morphological and functional aspects of evolution that followed the attainment of cellular organization.

Ways of origin of life

We have assumed that the processes leading to the origin of life and the origin itself took place on earth, and that it took place only once. However, for many centuries it was thought that life arises spontaneously and continuously from dead or decaying organic matter and even inorganic substances. Flies, for example, were supposed to arise from rotten meat. Experiments by Redi and Pasteur in the 1800s finally disposed of these ideas. Yet with the upsurge of interest in evolution at about the same time, the idea of *spontaneous generation* of life at least once in the remote past gained acceptance and continues to be believed today. We will return to this more or less orthodox concept later, but first we should look at some of the other prevalent thoughts about how life may have come to exist on earth.

One alternative to the spontaneous origin of life on earth relegates the whole process to an extraterrestrial source. Either life formed elsewhere and was brought to a well-formed earth subsequently, or life was part of the material that went into the earth at the time of its origin. These suggestions, of course, are possible, although the second is a bit difficult to conceive. The first puts the problem effectively out of reach of any currently possible direct method of study. It is true, however, that amino acids have been identified in some recent meteorites, indicating that some organic compounds can indeed form under extraterrestrial conditions.

An important aspect of the extraterrestrial line of thought is that life may have originated both on earth and elsewhere. This concept has enjoyed wide popularity and lies at the base of studies grouped under the general heading of *exobiology*. The basic premise of exobiology is that wherever physical conditions are suitable, life will come into being as a normal consequence of the evolution of matter through stages much like those presumed to have occurred on earth. Unfortunately, recent explorations of the solar system have materially reduced earlier hopes that some sort of life exists on the other planets. The possibility has not been completely ruled out, but what is currently known about physical conditions on the other planets makes it seem unlikely. If, however, planets are a normal phenomenon associated with sun-type stars, then there should be an immense number of planetary systems throughout the known universe, and at least some of these should have produced conditions suitable for life. We do know that within our own and other galaxies, elementary organic molecules have come into existence; the starting materials for life are present, it would seem, throughout the universe. The difficulty is finding a suitable environment in which these molecules can develop into a living system.

The alternative to some form of spontaneous generation is the creation of life by the intervention of a supernatural agent. Intuitive feelings concerning the nature of life and the apparent inexplicability of life by natural laws have led to this kind of interpretation. The idea takes two general forms. One is that life was created in the form of initial *living matter* and that the development of diversity was a natural consequence of this event. The other is that all living beings were created as they are and have not changed since. The first concept of supernatural creation is evolutionary; the second is not.

Approaches to the origin of life

Geology and biochemistry supply much of the data used in research on how life originated on earth. Although work in the two fields is largely independent, aims are the same and hypotheses cover data from both sources. Geologists have supplied organic compounds for chemical analysis and biochemists have posed questions that have directed studies of the rock record.

Geological studies include paleontological examination of extinct organisms preserved as fossils and of organic residues possibly left by organisms in the ancient sediments. They also include structural analyses of the rocks themselves to provide clues about the environment in those remote times. One of the most important contributions of geology is to determine as closely as possible the ages of the rocks. This involves measurement of amounts of radioactive minerals crystallized in the rocks at the time of their formation relative to the amounts of their decay products. The slow constant rate of radioactive disintegration provides a clock by which absolute age may be approximated. Despite the many difficulties involved in such analyses and the sizeable error to be expected from dates of very ancient rocks, the time scale so established is both useful and fairly reliable.

The principal focus of biochemical research has been upon present-day organic compounds derived from living organisms, from viruses, and from chemical syntheses in various experimental "atmospheres" and reactive environments. Studies have been made both of nonliving organic matter in the test tube (in *vitro*) and of living metabolizing organic matter in cells (in *vivo*). Experiments on living cells are made difficult by the fact that most of the analytical processes in current use interrupt or disturb the very processes they are attempting to elucidate.

A great deal of speculative thinking is central to both the in *vitro* and the in *vivo* approach; it is essential as a means to gain access to ill-defined problems by means of perceptive questions. Shakily supported hypotheses must be set up initially to provide a starting point for deductions which may be tested against empirical data. The general technique is to establish fairly comprehensive models of some major aspect of the origin of life. As a model is tested, it may be modified to fit emerging data and thus become more realistic. If the new data do serious violence to the model, both the data and the model are scrutinized carefully; usually it is the model which must be discarded. By these means, a closer and closer approach to a realistic understanding of events as they actually happened may be attained.

Geology and paleontology

Fossils and the time of origin

The earliest known remains generally acknowledged as organisms occur in the rocks of the Fig Tree group in Swaziland, South Africa (see fig. 3); their estimated age is 3.1 BY.[1] Within these sediments are silica deposits (*cherts*) containing extremely small spherules that have many of the features characteristic of bacteria and unicellular blue green algae (plate VI). A biogenetic origin of the spherules is indicated both on the basis of general form and of analysis of fine detail by electron microscopy. Organic residues from the surrounding rocks analyzed by gas chromatography affirm the presence of organic compounds of the sorts resulting from biogenic activities. From the beds of the Onverwacht group, in the same area as the Fig Tree, have come other remains of possible organic origin; these have yielded residues that suggest mixtures of both inorganically and organically derived hydrocarbons. Most paleontologists regard the spherules and organic traces as true fossils, but as always with such early and minute specimens, there is some element of doubt.

> **Figure 3** (opposite page) The geological calendar showing subdivisions and times in years. Eras for the Phanerozoic, the Paleozoic, Mesozoic and Cenozoic are indicated, and for these, the periods are entered. For the Proterozoic and Archaeozoic of the Precambrian time, only informal designations are used for subdivisions. Groups, series, and formations which carry fossils are listed for these times. Some of these are mentioned in the text. Dates in years are based on estimates from radioactive determinations. Partly after J.W. Schopf (1970).

---

[1] BY will be used to indicate one *billion years*. MY indicates one *million years*.

GEOLOGY AND PALEONTOLOGY

GENERAL TIME SCALE SHOWING SOME MAJOR SUBDIVISIONS
Groups and series in Precambrian are those with fossils.

	Eras	Periods	Ages in years
P H A N E R O Z O I C	CENOZOIC	Quaternary Tertiary	60 MY*
	MESOZOIC	Cretaceous Jurassic Triassic	230 MY
	PALEOZOIC	Permian Carboniferous Devonian Silurian Ordovician Cambrian	620 MY
	Informal divisions	Series and Groups	
P R O T E R O Z O I C	LATE PRECAMBRIAN	Pound Quartzite (0.6)** Brioverian Chert (0.6) Bitter Springs Formation (0.9) Skillogalee Dolomite Nonesuch Shale (1.0) Belt Group (1.1) Muhos Shale (1.3) Beck Spring Dolomite (1.3)	1.7 BY
	MIDDLE PRECAMBRIAN	Gunflint Formation (1.9) Valen Group (2.0) Transvaal Supergroup (2.0) Witwatersrand Supergroup (2.2)	2.5 BY
A R C H E O Z O I C	EARLY PRECAMBRIAN	Coutchiching Group (2.6) (no fossils) Soudan Iron Formation (2.7) Bulawayan Group (2.8) Fig Tree Group (3.1) Onverwacht Group (3.2)	3.5 BY
		No record of life	4.5 BY

*MY = millions of years; BY = billions of years (1000 million).
**This and the following designate the age (years ago) as x 10^9
Thus, 0.6 x 10^9 is 600 MY (600,000,000), and 2.3 x 10^9 is 2.3 BY (2,300,000,000).

We will take it as demonstrated that life existed on earth somewhat more than 3.0 BY ago. The total age of the earth is estimated to be between 4.5 and 5.0 BY. The oldest dated rocks are about 3.5 BY old, with possible dates to as much as 3.7 BY. If the earth's age is estimated at 5.0 BY, then there is an unrecorded interval of some 1.5 BY. The fossil remains in the Fig Tree group are somewhat diversified and are probably advanced to an unknown degree over the initial living organisms. Thus, it seems at least reasonable to extrapolate the origin of life back to 3.5 BY and possibly earlier.

The environment

What is known about the environment on the very early earth has been dependent on theories and evidence concerning the origin of the earth itself. Most of the data come from studies of ancient rocks and from the theoretical prerequisites for prebiological evolution prescribed by the biochemists. The earth's age (around 5.0 BY) has been estimated from studies of meteorites and from estimates of the age of the solar system. There have been numerous theories about how the earth originated, and these have changed through the years. There is no very great consensus even today, but certain things are agreed upon. Among these is the conclusion that both atmosphere and hydrosphere were present during the period when organic molecules accumulated and life originated.

The source of the first permanent atmosphere appears to have been the rocks of the earth in which gases were both trapped and generated during formation of the planetary mass. Whatever ephemeral atmosphere was present at the outset appears to have dissipated well before this time. The main gases for the first permanent atmosphere seem to have been the same substances that issue from volcanoes and hot springs today—carbon monoxide, carbon dioxide, hydrogen, nitrogen, sulfur, hydrochloric acid, and water vapor. Most models of the origin of organic molecules call for an atmosphere containing methane and ammonia as well, but the sources of these gases are not evident from the results of outgassing of rocks on the primitive earth.

The origin of the atmosphere has been correlated with the theory that there was a massive thermal event resulting in fusion of the earth not long before the time of the earliest dated rocks, about 3.5 BY ago. This, it has been suggested, is the reason that no older rocks have been found. However, the extremely complex geology of ancient rocks allows a chance that more ancient formations will yet be found, even though this seems unlikely at the present time. The important point here is that the inferred massive fusion would have resulted in massive degassing of the rocks, thus supplying the ingredients of the atmosphere. Only after this or a similar event could the sediments containing the first fossils have formed. The energy source for the fusion could have been radioactivity, gravitational readjustments, meteorite impacts, and tidal actions.

The thermal event has been tied in theory to the idea that the moon was "captured" by the earth rather than formed as part and parcel of an earth-moon system at the time of origin of the solar system. Though the actual time of capture is disputed, many scientists feel that it occurred at the time of massive fusion, thus augmenting gravitational and tidal energies. A number of interesting implications follow from this idea. The moon, at time of capture, may have been considerably closer to the earth than at present and may have had a different orbit. These conditions would have produced not only strong stresses in the solid earth, but also, once a hydrosphere had formed, would result in tides much greater in magnitude than those we see today. Massive reefs of algae formed between 2.0 and 1.0 BY ago suggest such tides. In addition, it is possible that the orbital distance of the moon from the earth resulted in shorter days. From much later, about 350 MY ago, has come evidence from growth lines of corals that days were in fact shorter at that time; that is, there were more days in a year than there are at present. Such environmental factors may have had important effects upon the origin of life. Too little is known, however, to do more than speculate about what these effects might have been.

The oldest known sedimentary rocks, those in the Onverwacht and Fig Tree groups, were deposited in water and show evidence that they were weathered on land prior to deposition. Clearly, there was both an atmosphere and a hydrosphere present, and perhaps the continents already had begun to form. The environmental conditions prevailing in the primitive atmosphere and seas can be inferred from looking at the chemical composition of the ancient rocks. The Onverwacht and Fig Tree contain various readily oxidized minerals in unoxidized form; this indicates *anaerobic* conditions of deposition, suggesting that there was no free oxygen in the atmosphere. There are carbonates present indicating the presence of carbon dioxide in the atmosphere and hydrosphere, but there is nothing to indicate the presence of ammonia. The fact that liquid water existed on the primitive earth tells us that, at least in some areas, the temperatures did not fall below 0° C, the freezing point, or climb to 100° C, the boiling point, of water.

It is important that oxygen was absent from the early atmosphere, for the organic compounds found in ancient rocks and those necessary for the origin of life could not have come into existence under oxidizing conditions. Without mediating enzymes, oxygen is poisonous to living matter. Also, without oxygen there could have been no ozone layer in the upper atmosphere to screen out the sun's ultraviolet radiation, and the potent rays could bathe the surface of the earth and oceans. The ultraviolet radiation could have provided some of the energy needed for organic synthesis, but could just as easily have acted to destroy large molecules of organic substance. Some scientists suggest the possibility of formation of significant amounts of oxygen by dissociation of water by the ultraviolet radiation, but this remains a matter of dispute.

Chemistry of the origin of life

Given some of the environmental limits, we may now come to grips with the matter of chemical events leading to the origin of life. Biochemistry is a complex field, and we can do no more than skim its surface, but we can examine the more important events and processes in a fairly general way. Unfortunately, direct evidence of the actual chemical events leading to the formation of the first living matter is not to be had. It is therefore necessary to develop models incorporating processes which meet the biochemical requirements and could have taken place in the kinds of environments indicated by physical analyses. The models that have been proposed vary greatly in detail, but in general they include the following steps. (See fig. 4 for a graphic portrayal of some of these steps.)

The elements

Step 1. The four principal elements that contribute to living matter are carbon, oxygen, hydrogen, and nitrogen. Iron and phosphorus are of critical importance in catalysis and energy-exchange reactions. Each of these elements is a common constituent of the earth, and presumably formed during the formation of the solar system. C,H,O, and N, and the compounds they form are the primary materials with which we are concerned.

Simple molecules

Step 2. Carbon monoxide (CO), carbon dioxide (CO_2), water (H_2O), methane (CH_4), ammonia (NH_3), and hydrogen and oxygen in molecular form (H_2 and O_2) are common molecules formed by the impor-

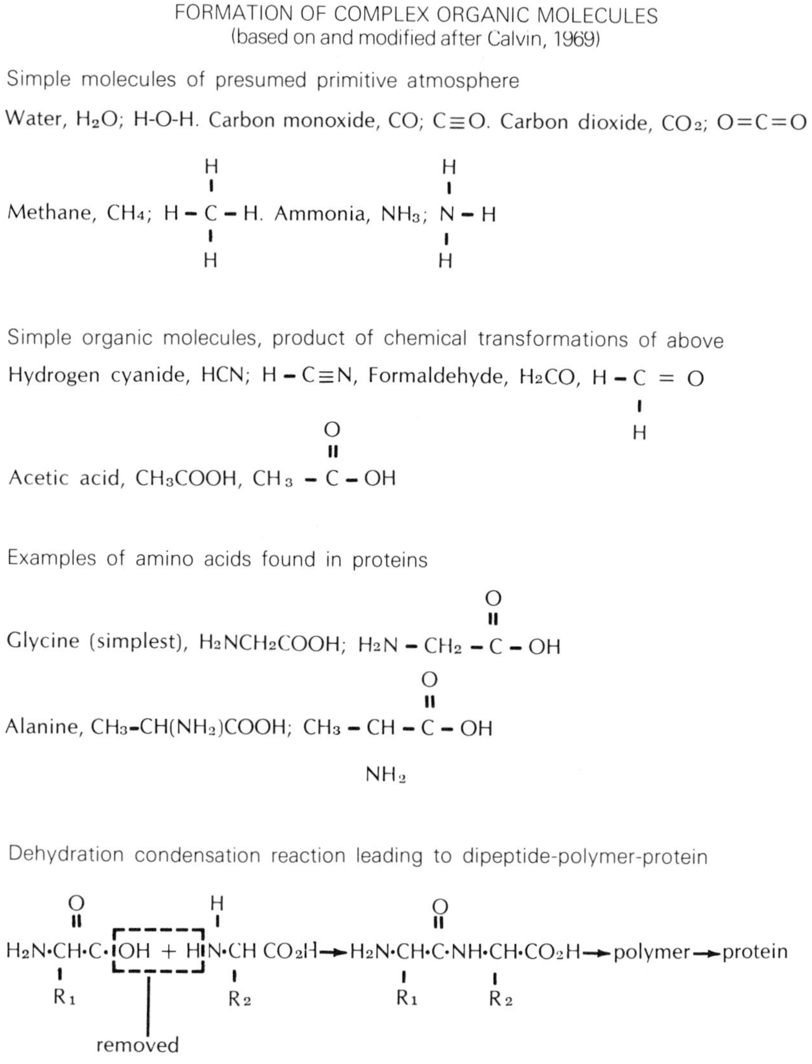

CHEMISTRY OF THE ORIGIN OF LIFE

FORMATION OF COMPLEX ORGANIC MOLECULES
(based on and modified after Calvin, 1969)

Simple molecules of presumed primitive atmosphere

Water, H_2O; H-O-H. Carbon monoxide, CO; $C\equiv O$. Carbon dioxide, CO_2; O=C=O

Methane, CH_4;
$$H-\underset{\underset{H}{|}}{\overset{\overset{H}{|}}{C}}-H$$
Ammonia, NH_3;
$$\underset{\underset{H}{|}}{\overset{\overset{H}{|}}{N}}-H$$

Simple organic molecules, product of chemical transformations of above

Hydrogen cyanide, HCN; $H-C\equiv N$, Formaldehyde, H_2CO,
$$H-\underset{\underset{H}{|}}{C}=O$$

Acetic acid, CH_3COOH,
$$CH_3-\overset{\overset{O}{\|}}{C}-OH$$

Examples of amino acids found in proteins

Glycine (simplest), H_2NCH_2COOH;
$$H_2N-CH_2-\overset{\overset{O}{\|}}{C}-OH$$

Alanine, CH_3-$CH(NH_2)COOH$;
$$CH_3-\underset{\underset{NH_2}{|}}{CH}-\overset{\overset{O}{\|}}{C}-OH$$

Dehydration condensation reaction leading to dipeptide-polymer-protein

$$\underset{\underset{R_1}{|}}{H_2N\cdot CH\cdot \overset{\overset{O}{\|}}{C}\cdot[OH}+\underset{\underset{R_2}{|}}{H|N\cdot CH\ CO_2H]}\rightarrow \underset{\underset{R_1}{|}}{H_2N\cdot CH\cdot \overset{\overset{O}{\|}}{C}\cdot NH\cdot }\underset{\underset{R_2}{|}}{CH\cdot CO_2H}\rightarrow \text{polymer}\rightarrow\text{protein}$$

removed

Figure 4 The pathway from simple organic molecules to polymeres by a series of steps as described in the text. Based on and modified after M. Calvin (1969).

tant elements. All of these, except for the oxygen, have been thought to be constituents of the primitive atmosphere. Most models depend heavily upon the presence of methane and ammonia, but the problem of the source of these molecules, discussed above, has yet to be resolved. Other models, in which these substances are not critical, have been proposed but are not as satisfactory.

Complex molecules

Step 3. Energy input into gaseous mixtures, including the molecules noted in step 2, produces an array of familiar larger molecules such as acetic acid, hydrocyanic acid, and formaldehyde. With additional energy input and, in some instances, catalysis, *amino acids* and other molecules associated with living systems can be formed. Amino acids are the common *metabolites* of all organisms and thus their production is of primary importance. Experiments involving energy input into various mixtures have determined that electric sparks, ultraviolet light, heat, and radioactivity all produce complex organic molecules. The end result is more dependent upon the composition of the *atmosphere* than it is on the nature of the energy input.

Development to this level could have proceeded by normal chemical processes which can be readily approximated in the laboratory. However, at some point, selection must be called into play to account for differential survival and concentration of certain molecules as opposed to others. Here, for the first time, we may picture an evolutionary process; that is, a process involving gradual directional change. The formation of the elementary molecules was a more or less random occurrence. Molecular selection started to operate when these molecules began to interact with one another to reach the most stable relative concentrations.

Cells today carry out their metabolic activities with the aid of efficient *catalysts*, the *enzymes*. Catalysts can lower temperature of reaction and accelerate reaction rates, and thus can favor increasing concentration of products of reactions in which they are operative. If a certain reaction product can catalyze the reaction that created it, the situation is called *autocatalysis*; the reaction is self-accelerating (see fig. 5). Such a system has a better chance of survival than others in which autocatalysis is not operative, because of its increased rate; molecular selection has occurred.

Monomers to polymers

Step 4. If we visualize a molecular mixture such as might have developed in an ancient sea, one in which a particular kind of monomer is concentrated enough that its molecules contact and react with one

CHEMISTRY OF THE ORIGIN OF LIFE

CATALYSIS AND AUTOCATALYSIS

$$A + B \xrightarrow{K} C + K \qquad A + B \xrightarrow{K_1} C + D \longrightarrow E + K_1$$

Autocatalysis Reflexive Catalysis

$$A + B \xrightarrow{K_2} C + K_1$$

$$P + Q \xrightarrow{K_1} R + k_2$$

Symbiotic Catalysis

Figure 5 Different forms of catalysis possibly important in the concentrations of organic molecules in the chemical evolution leading to the formation of cells. See the text for a discussion. Autocatalysis is essentially replication. Reflexive catalysis involves one or more additional reactions in production of a catalyst. Symbiotic catalysis is a possible mechanism whereby two reactions could gain selective advantage by taking place in close association.

another, we can predict the eventual formation of bimers and polymers of the substance. Thus, for example, amino acids might come to form proteins by the process of *dehydration concentration* (fig. 4).

Such reactions, of course, require energy input and, at the probable temperatures in the ancient seas, a catalyst. In cells, enzymes are very specific and efficient in their actions. In the prebiotic mix, however, catalysts were probably far less efficient but could act upon many different reactions. Dehydration synthesis can be accomplished by heating, by complex phosphates and by various compounds related to hydrogen cyanide. The cyanides have been proposed as probable agents under natural conditions, because they are usually abundant in artificially produced mixtures and because they can react well in an aqueous medium. Phosphorus, however, plays the key role in all energy-exchange reactions in modern cells; in the ancient seas it was available in the form of a mineral, calcium fluophosphate (*apatite*). Reactions involving phosphoric and hydrocyanic compounds offer possible ways in which dehydration synthesis may have occurred under primitive earth conditions.

We have been concerned specifically with the amino acid–protein polymerization, but the same general principles apply to the formation of such compounds as nucleic acids, lipids, and carbohydrates. Furthermore, they could have had autocatalytic or, more properly, reflexive catalytic properties. To this point, then, there are well-understood processes by which certain of the complex ingredients of living cells could be formed under presumed prebiotic conditions. Those molecules that had the selective advantage of autocatalysis would be the ones most likely to have survived and increased in concentration.

Association of different polymers

Step 5. Development of different suites of molecules using similar reactive pathways presumably would have resulted in a few predominant types of polymers. Among these, we may suppose, were proteins and nucleic acids. The latter had a capacity for fidelity in replication, and the former possessed multiple catalytic properties. The coupling of these two polymers to form interacting systems does occur, but the mechanism is far too complex to be profitably examined here. Suffice it to say that a *symbiotic* autocatalytic system may be developed in which each of the polymers forms catalysts or reactants for the reactions of the other. Once again, from an evolutionary viewpoint, this is of selective advantage.

Concentration and isolation

Step 6. Up to this point, even considering selection, the chemical processes themselves have been of the usual biochemical type; that is, they have statistical properties. The sequence is biochemically feasible. At a certain level of complexity, however, the chemistry begins to be ordered by the structure of the ingredients, and although this sort of ordered chemistry is characteristic of living matter and viruses, its method of development in nature is uncertain.

One important prerequisite for the development of higher levels of structural and chemical complexity is the concentration and isolation of the critical materials. The increased structural ordering is accompanied by increasing precision and control of chemical processes. At the present time, this must be considered the least-understood part of the process leading to the origin of cells, but there are a number of reasonable theories that deserve mention here.

One explanation, proposed and developed by the Russian scientist A.I. Oparin, is that organic molecules were first aggregated into cell-like bodies called *coacervates*. These bodies are minute spherules that form when solutions of macromolecules, usually proteins, are mixed. Each kind of macromolecule must be soluble in the same solvent (usually water), but must be relatively insoluble with one

CHEMISTRY OF THE ORIGIN OF LIFE

another. Thus, point separations occur in the mixture and droplets may form from these. Coacervates composed of multiple substances could have arisen on the way toward evolution of cells.

Bernal and some of his followers have shown that clay provides surfaces upon which concentration, ordering, and isolation of macromolecules can occur by *adsorption*. Conveniently, surfaces also tend to have catalytic properties, since they can bring reactants into juxtaposition. In the cell, the reactive surfaces are membranes.

Production of microspheres from substances termed *proteinoids* was accomplished by Fox. He heated amino acids to temperatures between 75° and 150° C and then cooled the mixture by contact with water. The resulting microspheres included polypeptides and seemed to demonstrate catalytic properties. Their method of formation suggests a method involving heat and evaporation by which isolated aggregates could have come into existence (fig. 6).

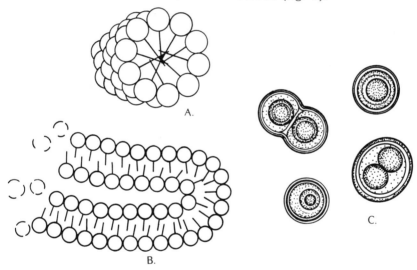

Figure 6 Micelles and protenoid forms of coacervates. *A*, A somewhat complex micelle representing an association of amphoteric molecules and leading to a bimolecular leaflet (See text). *B*, Protenoids formed by aggregates of amino acids, as described in the text. *A* adapted from S. Salthe (1972) and *B* drawn from photograph by S. Fox, in Salthe (1972).

The most intriguing idea yet proposed in this area is the *micelle* theory. There are certain kinds of molecules that have the property of being water soluble only at one end. Such molecules, called *ampho*-

terics, include soaps, detergents, and phospholipids. They will form a film at the surface of the water with their soluble ends down, or at higher concentrations will aggregate into globular micelles with the insoluble ends tucked into the center. At very high concentrations, the molecules will form long drawn-out bimolecular *leaflets* that have many of the structural properties of biological membranes, which also are constructed partly of phospholipids. At the temperature and concentrations at which these phenomena occur, the micelles and leaflets are very stable, and could offer resistance from destruction to an embryonic metabolic system. The membranelike structure of the leaflet could also act as a catalytic surface to molecules trapped within it. Most importantly, micelles have been constructed which contain enzymes and other compounds; the amphoteric nature of the molecules of the micelles shields and isolates these other molecules from the aqueous environment and its vicissitudes. The micelle theory is the most plausible one yet proposed for the concentration and "packaging" of early organic systems.

None of these processes has been demonstrated to produce living cells or even units that can readily be considered as forerunners of cells; they merely illustrate ways in which isolated aggregates of organic molecules may form. Each theory has its strong and weak points, but none can be isolated and shown to be the process that actually occurred. Indeed, perhaps the origin of cells involved combinations of these processes or incorporated a method of aggregation about which we have no knowledge.

An evolutionary model

We will now follow through a model of how life may have arisen on earth. It conforms with both biochemical and geological information, but there are alternatives at all levels, and the sequence of events itself may be in error.

During the interval of chemical evolution leading to living systems, the earth had land masses, bodies of standing water, a reducing (as opposed to oxidizing) atmosphere, and temperatures probably not differing too sharply from those at the present time. Energy was available in the form of radioactivity, electric sparks, ultraviolet light, and heat, and this energy acted upon gases in the atmosphere to produce simple organic molecules. These molecules, presumably including amino acids, accumulated in the waters of the earth to form a dilute organic *soup* whose concentration gradually increased with time.

AN EVOLUTIONARY MODEL

Particular kinds of reactive molecules came into contact with one another and, perhaps with the aid of relatively inefficient catalysts, reactions occurred. Those molecules having some selective advantage persisted and increased proportionately, more reactions occurred, and larger, more complex molecules were formed. Continuing concentrations of selected large organic molecules led to reduced chemical diversity in the soup. Very probably, symbiotic relationships between polypeptides and polynucleotides developed, involving selectively advantageous symbiotic autocatalysis.

Purely chemical or molecular natural selection operated to this point. In contrast to Neodarwinian selection, it involved only the replicating mechanisms (reactions) themselves. In a crude analogy, the total array of substances in the organic soup might be thought of as equivalent to the *genetic pool* of a species. In Neodarwinian evolution, natural selection acts upon the *product* of the gene pool, while in chemical evolution it acts directly upon the pool itself. With the development of symbiotic autocatalysis, the first evidence of a product, the *phenotype*, may be seen. Survival at this stage was directly related to molecular concentration; eventually, concentrations must have increased to the point that aggregation and finally individualization could take place.

One of the pressing problems in prebiological evolution concerns how to keep the newly-forming organic molecules out of the reach of the highly destructive untraviolet radiation pouring down onto the earth in the absence of an upper atmospheric ozone layer. Several models have been proposed. Among the best of these is a model which also provides a means of concentrating the organic molecules. In the model, the organic soup is envisaged as existing in relatively small closed pools with a gentle circulation. Heavy molecules could sink deeper down into the pools, where the overlying water could shield them from destructive radiation. Light of less destructive wavelengths, however, could penetrate to the molecules, providing them with a potentially usable energy source. This situation might set the stage for coacervate, microsphere, or micelle formation, depending on temperature and concentration of the soup.

The seashore, with its estuaries, beaches, and tide pools, offers various possibilities for concentration as a result of evaporation in the presence of marked temperature fluctuations in temporary pools and lagoons. Coacervates, proteinoid spherules, and micelles could have formed under these conditions. Concentration by adsorption on clays, which could have been present near shorelines, might have occurred and would have offered some protection against radiation.

These are all valid possibilities and there are still others. Each has been proposed to meet the complex requirements indicated by the evidence that is available and to harmonize the widely differing bodies of data gathered by biochemists and geologists. Each theory must be considered as a model to be tested and not as a final answer to the full array of problems surrounding prebiological evolution.

We do know that cells developed on earth, or at least appeared there, about 3.5 BY ago. At that critical time, prebiological evolution came to an end, and Neodarwinian selection began to exert its powerful directional influence on the first real organisms to appear on the face of the earth.

IMPORTANT CONCEPTS

AMINO ACID: monomers with the general chemical formula:

$$R-\underset{\underset{NH_2}{|}}{\overset{\overset{H}{|}}{C}}-COOH$$

The R stands for various organic groups, and it is in the composition and structure of R that amino acids differ.

AUTOCATALYSIS: (See fig. 5). A chemical reaction in which the catalytic agent is a product of the reaction (see *catalysis*). Reflexive autocatalysis occurs if the product of a reaction, which is part of a sequence, catalyzes an earlier reaction in the sequence.

BY: billion years.

CARBOHYDRATE: a sugar or sugar derivative. Single sugars are called *monosaccharides* and sugar polymers are *polysaccharides*. Monosaccharides have a special significance and form part of the structure of *nucleotides*.

CATALYSIS: the cause or acceleration of a chemical reaction by a substance (*catalyst*) that is not itself permanently affected by the reaction. It decreases the activation energy needed for the reaction to take place. *Enzymes* (*proteins*) are critically important as catalysts of chemical reactions of cells.

DEHYDRATION CONCENTRATION: the formation of chains of molecules (*polymerization*) by bond formation involving the loss of water molecules (see fig. 4).

ELEMENT: one of the class of substances that cannot be separated into other substances by usual chemical methods. We will deal largely with the following elements, given with their symbols.

Oxygen, O, molecular form O_2 Iron, Fe
Hydrogen, H, molecular form H_2 Phosphorus, P
Nitrogen, N, molecular form N_2 Sulfur, S
Carbon, C

ENZYME: proteins which are highly efficient and selective organic catalysts. They have specific actions and are named by the substrate they act upon plus the suffix, -ase. Thus, an enzyme acting on carbohydrates would be called carbohydrase, and so on. Some enzymes are extremely specific in their actions and others less so.

LIPIDS: fats, oils, waxes. Insoluble in water. Under hydrolysis (addition of water), lipids form fatty acids and glycerol (glycerine).

METABOLISM: chemical activities of living matter. *Metabolites* are the substances acted upon or produced by metabolism.

MOLECULES: the smallest physical unit of an element or compound. The first case includes two or more like atoms and the second two or more different atoms.
Example: methane, with 1 carbon atom and 4 hydrogen atoms:

$$CH_4 \quad \text{or} \quad \begin{array}{c} H \\ | \\ H-C-H \\ | \\ H \end{array}$$

Macromolecules are those of very high molecular weight, very long chain polymers, such as nucleic acids of proteins.

MONOMER: simple compounds that form links in the chains termed *polymers*. Specifically, a molecule of low molecular weight capable of reacting with similar or different molecules to form polymers. Examples are molecules of an amino acid or of a nucleotide.

MY: million years.

NUCLEOTIDE: (also *polynucleotide, nucleic acid*). Nucleotides are the building blocks of nucleic acids. Each has three constituent parts—a sugar, a nitrogen base, and a phosphate group. In DNA (deoxyribonucleic acid), for example, the nucleotides have a 5-carbon sugar (deoxyribose) base as follows:

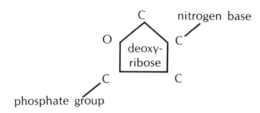

The phosphate base remains the same in all, but there are four different nitrogen bases which are critical in the formation of DNA. RNA (ribonucleic acid) is formed of nucleotides with a different sugar base (ribose) and one different nitrogen base.

PHOTOSYNTHESIS: the formation of organic compounds in cells utilizing the energy of light.

POLYMER: a long molecular chain composed of multiple monomers Specifically, molecules of high molecular weight (macromolecules) formed by dehydration condensation of many small molecules.

POLYMERIZATION: the process of development of polymers from monomers. In living cells, accomplished by dehydration condensation.

PROTEIN: a macromolecule which is a chain of *polypeptides* (polypeptides being a short chain of amino acids).

REFERENCES

Calvin, M. 1969. *Chemical evolution.* New York and London: Oxford University Press. p. 278.

Cloud, P.E., Jr. 1968. Pre-metazoan evolution and the origins of the metazoa. In *Evolution and Environment,* edited by E.T. Drake. New Haven and London: Yale University Press. p. 72

Cloud, P.E., Jr., ed. 1970. Adventures in earth history. San Francisco: W.H. Freeman & Co. p. 992.

Fox, S. 1964. *Origin of prebiological systems and their molecular matrices.* New York: Academic Press Inc. especially pp. 361–73.

Keosian, J. 1964. *The origin of life.* New York: Reinhold Publishing Corp. p. 118.

Orgel, L.E. 1973. *The origins of life.* New York: John Wiley & Sons, Inc. p. 237.

Salthe, S. *Evolutionary biology.* New York: Holt, Rinehart and Winston, Inc. pp. 35, 32.

Schopf, J.W. 1970. Precambrian micro-organisms and evolutionary events prior to the origin of vascular plants. *Biological Reviews* 45: 319–52.

Shklovskii, I.S. and Sagan, C. 1966. *Intelligent life in the universe.* New York: Dell Publishing Co., Inc., Delta. p. 509.

4

The early evolution of life

Perspectives

In this chapter, we will consider evolution at the cellular level of organization, including those multicellular organisms that lack differentiated tissues and organ systems, and both procaryotic and eucaryotic cell types. The organisms we will be dealing with often are lumped into the kingdom *protista* in its broadest and most inclusive sense. However, it must be continually borne in mind that structurally, functionally, and evolutionarily, the Protista forms an extremely heterogeneous array. This diversity has led many biologists to divide the Protista into two kingdoms. One, Kingdom *Monera,* includes one-celled organisms with very simple cells, the *procaryotes;* the other, Kingdom *Protista,* is restricted to one-celled organisms with cells which have a well-formed nucleus, the *eucaryotes.*

Hopefully, we will gain a fair idea of the general course of evolutionary history at the cellular level from the considerations dealt with in this chapter. Also, and more importantly, we will have the opportunity to examine a number of interesting evolutionary phenomena, many of which are unlike those characteristic of the better-known evolutionary history of *Metazoa* and *Metaphyta.* Some of the events in the very early history of life do fit the dogma or paradigm of Darwinian natural selection nicely, but, as we shall see, some do not fit the pattern at all. It is these evolutionary idiosyncrasies that will be taken up in the latter parts of the chapter.

Evolutionary landmarks

Two of the most crucial events in the development of living systems were the origin of cells about 3.5 BY ago, marking the beginning of Darwinian selection, and the origin of complex multicellular organization in plants and animals around 700 MY ago. Between these two events, about 70% of the history of life and 50% of the duration of the earth itself elapsed. In this entire vast span of time, the Monera and Protista were the only existing organisms, and because of our tendency to lump many diverse organisms into single categories which we then think of as evolutionary units, there is a great temptation to look upon this immense block of time as a period when changes were slow and relatively little evolution was occurring.

The facts of the case argue quite to the contrary. We know that at the end of the long interval there was an incredible diversity of one-celled organisms which included, among other things, the lines that would eventually lead to higher plants and animals. We also know that a number of evolutionary milestones were passed, among which were:

1. the origin of photosynthesis,
2. the origin of *aerobes* from *anaerobes,*
3. the origin of mitotic cell divisions,
4. the origin of meiosis and sexual reproduction,
5. the origin of cell aggregates, including colonial organisms and *thallophytes,* and the formation of *bioherms,*
6. the origin of eucaryotic cells.

Each of these events poses its own set of questions and problems, including the order of appearance, events leading to its origin, and its environmental and evolutionary implications. However, they are in many ways interdependent; despite the relatively sparse data available for any of them, there is enough information from our knowledge of processes and structures as they exist today to propose some

OVERVIEW OF THE GEOLOGICAL RECORD

far-reaching and thought-provoking theories. The heyday of evolution of one-celled organisms saw some of the most profound evolutionary events in the history of life. All of the innovations have since become incorporated in some way into organization at the multicellular level. The rest of the chapter will be devoted to close examination of these innovations and their attending theories. The metazoans, in all their familiar glory, will be taken up in detail later on.

Overview of the geological record

The geological history of the earth has been subdivided into major units based on prominent biological events shown by the fossil record and physical discontinuities and sequences in the rocks themselves. The general scheme is outlined in figure 3, chapter 3. The earliest period in the Phanerozoic is the *Cambrian*; this was the time when rocks containing rich deposits of metazoans first appeared. This period's extreme importance in the earth's history, based on this fact, is shown by the common designation of all earlier rocks as *Precambrian*, an informal but very useful term. For convenience, we will use this term along with the informal subdivisions of *early*, *middle* and *late* Precambrian, rather than the more technical subdivisions of Proterozoic and Archaeozoic (Archaean), shown in the chart in figure 3. The chart includes names of the rock units that contain most of the important fossil remains from the Precambrian. Altogether, for this interval of some 2.5 BY, only about 35 sites have been found in which identifiable fossil remains are present. The fossils are preserved in fine-grained silica (*chert*) and represent the products of unusual conditions of deposition. Most of the organisms alive at that time were not preserved, and there undoubtedly were many different kinds of creatures that left no record of their existence at all.

The available record for the Precambrian, then, is not awfully impressive. Although it improves somewhat from older to younger beds, it is always spotty, incomplete, and biased, and any conclusions based on it must be somewhat suspect. Still, it is immensely better than what was available as little as 20 years ago. With the amount of investigation going on in Precambrian paleontology today, the record will continue to improve. Both fossil organisms and organic residues can yield considerable information. The sum is much less than what would be desirable, but a great deal can be learned from what we do have.

Data and the main events of history

Early Precambrian

The earliest known fossil organisms are unicellular procaryotes, the *Monera*. Comparisons with living forms and the conditions of occurrence suggest that they were *anaerobic heterotrophs* that presumably lived on *abiogenic* matter, organic molecules not formed by life processes. The procaryotes, bacteria, and blue green algae, appear to have been present in abundance in these earliest beds. If the identification of the blue green algae is correct, then it is possible that oxygen-producing *autotrophs* had already developed. However, it seems more probable that, although some sort of photosynthesis was taking place, free oxygen was not liberated by the process.

It is singularly striking that the very early life forms should resemble living organisms so closely that they may be placed with some confidence in the same taxonomic families. This is true not only for the individual unicells, but for larger biogenic structures called *stromatolites*, which were composed of layers of inorganic material deposited by blue green algae. Similar stromatolites exist in limited distribution today; they form reefs that provide microenvironments for other organisms. Stromatolites occur in the Bulawayan and Soudan Iron Formations of Rhodesia, some 2.8 BY old. These formations provide evidence that oxygen-producing, autotrophic, photosynthetic blue green algae were present and that free oxygen was being supplied to the environment.

Exactly how much oxygen was produced at that time and later is somewhat uncertain. We do know that in the time between the deposition of the Soudan Iron Formation and the end of the middle Precambrian, highly oxidized red iron formations were developed. Algae are found associated with the red beds, and presumably the oxygen released by these and the activity of iron-oxidizing bacteria were responsible for the oxidation of such extensive quantities of iron. It is not unlikely that, by this time, there was some small amount of free atmospheric oxygen as well.

Middle Precambrian

There is much more fossil evidence available from the middle Precambrian than there is from earlier rocks. In the Gunflint Iron Formation, about a dozen different kinds of organisms have been found; most, possibly all, of these are blue green algae. Interestingly, there

DATA AND THE MAIN EVENTS OF HISTORY

are some filamentous chains of cells as well as solitary unicells present (see Plate VI). *Chemosynthesizing* bacteria were in existence at this time, and algal stromatolites were widespread around the earth in shallow coastal waters.

There is little question that organisms were contributing oxygen to the atmosphere by the middle Precambrian, adding to whatever oxygen was produced by dissociation of water by ultraviolet light. If the dissociation process were a significantly important source of oxygen, then appreciable amounts of the substance might have existed in the atmosphere for a long time before organisms began to contribute their share. Oxygen availability is critical for two reasons. First, it has a great deal to do with the energy available for organic synthesis by ultraviolet light; oxygen, in the form of ozone (O_3) is an effective screen for ultraviolet rays. Secondly, the development of oxygen-metabolizing or even oxygen-tolerating organisms could not have developed until a steady source of oxygen in the seas and in the atmosphere was present. Most of the evidence indicates that such an atmosphere did not develop until the middle Precambrian. Conveniently, this atmosphere formation coincides with what we consider to be the correct theoretical sequence of events in the evolution of metabolic systems in procaryotes.

Late Precambrian

During this time, blue green algae were ubiquitous, and many of the forms are indistinguishable from living genera. Various kinds of bacteria have been recorded, among them the ancestors of modern types which they resemble closely. Most significant, however, is the presence of the first recorded eucaryotic cells, Protista. From the Bitter Springs formation (see fig. 3) have come cells with included concentrations of organic matter which are easily interpretable as the remains of nuclei, plastids, and mitochondria. Furthermore, some of these cells have been caught in the act of reproducing; and there is evidence both of *mitosis* (two cells adjoining) and *meiosis* (tetrads of cells). If meiosis truly was occurring at this time, sexual reproduction had already developed.

Most of the kinds of eucaryotic algae—green, red, and possibly brown—have been found in beds of late Precambrian age. Fungi, too, have been identified from these beds. Eucaryotes first appear in rocks about 1.3 BY old, and it seems probable that this is very close to the time of their origin. The earlier finds that have been reported are extremely uncertain.

In the latest Precambrian, a few metazoans occur and multicellular plants are well represented by algae which are quite similar to some living today. Shortly thereafter, with the beginning of the Cambrian, complex metazoans of several kinds appear in abundance. One of the major puzzles of the history of life as seen in the fossil record is this apparently sudden development of a diversity of advanced animals.

Evolution: the nature of events

In some respects, this section will be one of the most difficult to follow, but it is also one of the most important for a full understanding of the evolution of life. The content is intrinsically difficult, relating detailed biochemistry, biology, and paleontology. It will be kept as simple as possible without glossing over the most critical events.

The primordial cell

Evolution is often spoken of as proceeding *from amoeba to man*. This is fairly accurate as a portrayal of only one line of development, but there are many others. The amoeba in this case represents the typical *primordial cell*, a cell probably enough like a living amoeba to be called at least *amoeboid*. A primordial cell capable of self-duplication must have had at least the following components:

1. DNA (deoxyribose nucleic acid)
2. RNA (ribose nucleic acid)
3. Amino acids, proteins, peptides, enzymes
4. Sugars
5. Cytoplasm (a watery substance containing salts and metabolites)
6. Cell membrane

Nothing of this skeletonlike simplicity exists today, and there is no fossil record of such a cell. It is only a theoretical construct, albeit a necessary one. Presumably, this primordial "first cell" was a heterotroph, feeding on abiotically formed organic matter.

Procaryotic cells

As noted in our brief section on the geological record, only procaryotic cells were present between 3.5 BY and 1.3 BY ago. Today, the procaryotes are represented by a wide variety of bacteria and blue green algae, many of which seem to date back to the Precambrian with little morphological, but probably much biochemical, change. Thus, a gross look at the record suggests that not much evolution was going on during the 2.0+ BY of exclusively procaryotic deployment.

Procaryotes reproduce predominantly by simple *binary fission*; the genetic material duplicates, and the cell simply splits in two. None of the complex mechanisms of mitosis or meiosis are present. Some recombination of DNA between individual bacteria does occur, but for the most part, genetic changes in a cell are transmitted only to its direct descendants. There is little spread of variation through populations of procaryotes. For this reason, rates of evolutionary change probably were slow.

Even considering snail's pace rates of evolution, two billion years is a long time, and many things could and did happen. Several fundamental organizational changes of the primordial cell took place, mostly involving the ways in which energy sources were utilized and the origin of new metabolic pathways and processes. The actual structural and chemical modifications need not have been great to have a profound effect on resource utilization, and the use of new resources alone could have opened up a broad horizon of new ways of life. Exploitation of the niches within such a new environmental zone is called *adaptive radiation*.

The first living cells probably were heterotrophs, feeding upon nonbiogenic organic molecules floating free as part of the organic soup. There are primitive heterotrophs alive today which probably resemble these early forms. The difference is that modern forms no longer are surrounded by a nutrient soup; they must feed upon organically produced molecular matter. Heterotrophic procaryotes have undergone a fair amount of adaptive radiation, but their fundamental patterns of organization have not changed radically. For example, some bacteria developed the capacity to *fix* various atmospheric chemical substances such as carbon dioxide and nitrogen. Others, the chemosynthesizers, gained the ability to oxidize iron, sulfur, and other elements. Some of the heterotrophs are aerobes, others are anaerobes. The differences among them, however, are mostly biochemical.

One of the major steps in procaryotic evolution was the development of photosynthesis, which opened up new possibilities of energy utilization. Both bacteria and blue green algae evolved photosynthetic pathways, but there are differences; the bacteria do not produce oxygen from their process, while the blue green algae do. The relatively simple biochemical change leading to biosynthesis of chlorophyll and oxygen production was one of the most important steps in all of evolution. The blue green algae, successful organisms in their own right, opened the way for the evolution of the higher plants and, indirectly, the evolution of higher animals which could utilize the plants for food.

Each of the several major lines of procaryotic evolution, except the most persistently primitive, made the shift from anaerobic to aerobic life, attaining the ability to survive in the presence of oxygen, even if they could not metabolize it. The appearance of the same phenomenon in several taxonomic groups is not an uncommon occurrence; we shall see other examples of such evolutionary convergence among the metazoans and metaphytes.

Before we take our leave of the procaryotes, one other matter, involving the blue green algae, must be noted because of its evolutionary significance. It seems that once the basic blue green algal structure was established, and this was way back in the Precambrian, there was little subsequent change. This means that they did not evolve significantly in this way for 2.5 BY, quite an accomplishment in a world of constantly changing environmental conditions and fluctuating populations. Nonevolution, or very slow rates of change, also occur in other groups, but to a much lesser degree. Nonevolution is often cited as a fact that evolutionary theory cannot explain, but this comes basically from a misunderstanding of the nature of the theory.

There are two types of adaptations—one is *phylogenetic*, that is, evolutionary, and involves a permanent change; the other is *temporary* and involves moment-to-moment changes in physiology to keep the organism in equilibrium relative to its environment. Organisms with broad environmental tolerances are particularly adept at physiological adaptation, and can shunt their life processes through several different metabolic pathways, depending upon prevailing conditions. The effects of natural selection are, of course, minimized under such circumstances.

The blue green algae have specialized toward efficient physiological adaptation. Their biochemical processes and responses to change are so plastic that cells may take on different physical appearances in different environments. Although more study is needed, it appears that at least part of their capacity to adjust is under genetic control, so that upon exposure of the organisms to different temperatures, salinities, light intensities, and so forth, different parts of the genetic potential are expressed in the adults. Thus, they have a very special adaptation—the capacity to adjust—which, carried to the extreme, results in essentially zero evolution.

Eucaryotic cells

All organisms except the bacteria and blue green algae are eucaryotes, whether they are single- or multicelled. The distinction between eucaryotes and procaryotes is so complete that every organism unequivocally can be called one or the other; neither the modern biota nor the fossil record provides any clue to a transition. The usual interpretation, nonetheless, is that the standard evolutionary processes—mutation, recombination, and gradual accumulation of small changes—were in operation during the origin or eucaryotes.

The standard interpretation is not impossible, for by no means have all kinds of organisms survived, and the intermediates simply may not have been preserved. Evolutionists, rightly, are reluctant to abandon an applicable and useful way of looking at the changing organic world, unless the evidence contrary to it is too compelling to be comfortably brushed aside. Thus, it is assumed that the eucaryotes evolved according to the standard formula and that the eucaryotic pattern arose *only once*. Any major change that leads to a new major group is considered to have happened but once, mainly because it seems unlikely that even parallel changes in different lines would have been insufficiently alike to produce the unity of structure found in the new group. Stated otherwise, all major groups have a single origin, they are *monophyletic*.

In the case of the procaryote-eucaryote relationship, it is becoming increasingly evident, especially considering the great strides made in biochemistry in recent years, that the usual explanation does not fit all the facts. Perhaps this is the result of inadequacy of the facts or of our ability to interpret them, but this seems less and less likely. A brief look at an alternative which may have great evolutionary significance is worthwhile here. It must be remembered that this theory is not the only way, or even the most popular way, to view eucaryotic origins, but it is emphasized here because it offers a unique solution to one of the most perplexing problems in the history of life.

The alternative proposal is that the eucaryotes arose as the product of *symbiosis*, the cooperative association of two or more separate organisms. It has been studied in depth by Lynn Margulis (1970), and what is included here owes its base to her work. For simplicity, we will assume that it is the correct interpretation, although this may or may not be the case. Thus we can, without interruption, develop the theory of symbiotic origins without digressions into complex evidence to the contrary. The presentation of the full case, with its supporting evidence, is equally complicated, and we will be able to touch only upon the high spots.

There are many clear-cut differences between procaryotes and eucaryotes. The ones that will be important to our purposes are illustrated in the table. Any explanation of the origin of eucaryotes must

Procaryotes	Eucaryotes
1. Mostly small—1–10 microns.	1. Mostly large—10–100 microns.
2. No nuclear membrane; circular chromosomes.	2. Membrane-bound nucleus; linear chromosomes.
3. Cell division mostly by binary fission; sexual reproduction in the form of transformation, transduction, etc.	3. Cell division by mitosis; sexual systems present in the form of meiosis with the production of gametes and zygotes.
4. Simple bacterial flagella, if flagellate.	4. Flagella and cilia, when present, complex, with a 9 + 2 pattern.
5. Membrane-bound organelles absent.	5. Organelles (mitochondria, plastids in photosynthetic plants, etc.) are present and membrane-bound.

account for these fundamental differences. All of the innovations are significant, for they have produced a unique and evolutionarily stable operating system with standard, highly efficient metabolic features. From an evolutionary point of view, it is the operation of these various new processes in the development of sexual reproduction that is most significant. Once this type of reproduction had arisen, the exchange of genetic material between individuals was realized, and the spread of variation through populations took place. Under these conditions, Darwinian selection can operate at maximum efficiency to produce permanent directed evolutionary change. We will take a closer look at both mitosis and meiosis later on in the chapter.

Figure 7, modified after Margulis, presents a very simple interpretation on the steps in the symbiotic model of evolution of the eucaryotic cell. It will be noted that several steps are involved and that most of these involve a symbiotic association. In the first step, an anaerobic microbial host harbored an aerobic bacterium with result-

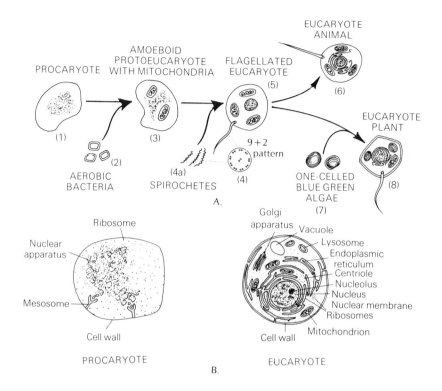

Figure 7 The suggested course of evolution from procaryote to eucaryote cells by symbiosis. Numbers 1–8 are referred to in the discussion in the text. Adapted from L. Margulis (1970).

ing augmentation of the metabolic capacities of the entire system. The bacterial component specialized to form *mitochondria*, and the host provided *cytoplasm* and a *nucleus*. This produced the *amoeboid* (3). It is interesting to note that in modern eucaryotes, mitochondria have their own circular DNA and can replicate independently of the nucleus.

From the amoeboid arose *amoebae*, some of which have survived to the present. Most, however, have been replaced by *amoeboflagellates* (5) which were formed by a second symbiosis involving *spirochaetelike organisms* (4a). In cross section, spirochaetes show a 9 + 2 arrangement of rodlike *fibroproteins*; this arrangement is characteristic also of eucaryotic flagellae, cilia, and centrosomes, the last of which are intimately involved with mitotic and meiotic processes. Thus, by a complex series of steps, the spirochaete symbiont adjusted to the reproductive system of its host and, combining with it, gave rise to mitotic cell division. The flagellar part of the spirochaete offered a new suite of locomotory potentials to the ecuaryotic cells, and thus opened up new ways of life that have resulted in extensive adaptive radiations. Modern protozoans are actually classified according to their means of locomotion.

From the amoebo-flagellates (5) came protozoan animals (6) and eucaryotic plants (8). The origin of the eucaryotic plants involved yet another symbiosis, this time with the amoebo-flagellate playing host to a blue green alga; the alga supplied its photosynthetic capability. Its host contributed its nucleus, cytoplasm mitochondria, and spirochaete-derived structures. *Chloroplasts*, like mitochondria, can replicate themselves and contain circular DNA. Gradually, the metabolic processes of the plastids and their hosts became coupled, and the result was a photosynthesizing eucaryotic organism.

There is much evidence for this postulated sequence of events, but it raises several troublesome questions. Its various steps are based mainly upon studies of recent organisms some billion and a half years remote from the actual occurrence. Furthermore, this hypothesis breaks away from a tested, acceptable interpretation of evolutionary processes. Other attempts have been made to explain the origin of major groups through symbiotic associations, but most of these, with good reason, have been rejected.

Two critical matters are not well displayed in the diagram in figure 7. One is that the process of symbiosis probably did not occur only once; the likelihood is great that it was a commonplace *experiment* with multiple origins of somewhat similar cells, but incorporating different kinds of algae or bacteria. The different photosynthetic pigments found in different groups of plants offer evidence for a multiple, or *polyphyletic*, origin of the *metaphytes*.

The other matter concerns the origin of mitosis and then meiosis as an elaboration and extension of it. The symbiotic hypothesis envisages mitosis developing in a complex series of steps as the *seed* of the flagellum coupled its reproductive capacities with those of the host nucleus. This seed, today called a *centrosome*, also has self-reproductive powers; but unlike the mitochondrion and plastid, it reproduces at the same time as the host nucleus and actually plays an integral part in the reproduction of the entire cell. The process of mitosis and the related process of meiosis are important to an understanding of evolutionary processes in higher organisms.

Mitosis, meiosis, and the evolution of life cycles

In most of the procaryotes, cell division is a relatively simple process. The genetic material, carried on a single circular chromosome,

EVOLUTION: THE NATURE OF EVENTS

duplicates itself, and then the entire cell divides into two new cells, each with a single circular chromosome. The mitotic eucaryotes achieve the same end, but the presence of a possibly spirochaete-derived self-replicating centrosome and numerous linear chromosomes tend to complicate matters somewhat. Very briefly, the events of mitosis are as follows (see fig. 8):

1. A metabolic, synthetic phase during which cell growth occurs prior to reproduction (**Interphase**)
2. A replicative phase where each linear chromosome duplicates but remains physically attached to its sister (*sister chromatid*). At the same time, the centrosome outside the nucleus divides to form two centrioles which migrate to opposite sides of the nucleus but remain connected by protein fibers, collectively called the *spindle apparatus*. In preparation for the "escape" of its enclosed chromosomes, the nuclear membrane begins to break down. (**Prophase**)
3. An organizational phase, including the attachment of each pair of sister chromatids to the center of one of the spindle fibers. (**Metaphase**)
4. A kinetic phase in which the attached sister chromatids break apart and migrate toward opposite centrioles on the spindle fibers. (**Anaphase**)
5. A restorative phase in which the cell membrane pinches off to form two new cells, and the nuclear membranes reform. (**Telophase**)

During interphase, which in most cells is of much greater duration than all of the other phases put together, the organism is effectively *haploid*—it has only one set of chromosomes. From prophase to the separation of the two new daughter cells, there are two sets of chromosomes present as a result of duplication of the original set. However, this *diploid* phase is short-lived so that the entire life cycle of the organism may be said to be *haploid-dominant*. All eucaryotic cells undergoing mitosis, including the remote ancestors of higher plants and animals, had haploid-dominant life cycles.

In sexually reproducing organisms, additional complexity has been added to the process. The critical event is the *fusion* of two haploid (1N) cells, the *gametes,* to form a single diploid (2N) zygote. This cell contains two sets of *homologous* chromosomes, one set from each gamete, with each chromosome having a homologue that is identical to it in gross structure and which carries gene sequences for the same

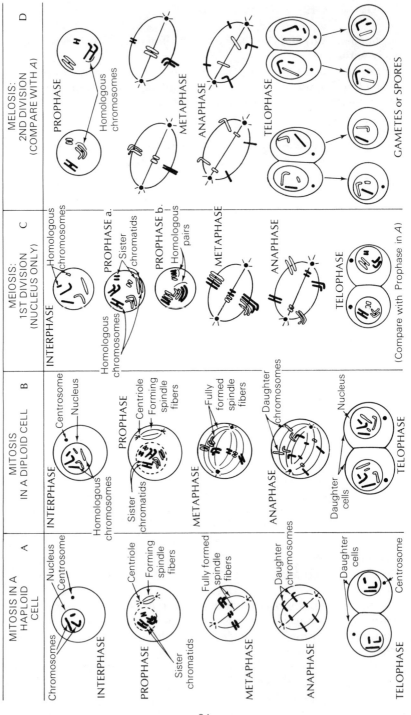

characters. In such a cell, there is emphasis on both the haploid and diploid stages.

An important consequence of arriving at the 2N state by means of *syngamy* (fusion of gametes) is that when chromosome replication occurs in preparation for cell division, the cell will have four sets of chromosomes. At this stage, there are two possible alternatives. The simpler and more common one is for the cell to undergo mitosis, separate the sister chromatids formed from the chromosome replication, and produce two 2N daughter cells (see fig. 9). This sort of mitosis is responsible for the multiplication of cells from the zygote to form an adult plant or animal composed of many 2N cells.

The other alternative is meiosis, which involves two separate cell divisions. It proceeds in the following way (see fig. 8):

1. *First meiotic division:* separation of homologous chromosomes.
    a. Each pair of sister chromatids lines up next to its homologue and both attach to the spindle apparatus.
    b. Cell division proceeds, and each pair of chromatids is separated from its homologue as they move toward opposite centrioles.
    c. Two diploid cells result.
2. *Second meiotic division:* separation of sister chromatids (Mitosis)
    a. The sister chromatids, still attached to one another, line up on the spindle fibers.
    b. Cell division occurs, and the sister chromatids are separated as they move toward opposite centrioles.
    c. Each of the two cells produced by the first meiotic division gives rise to two haploid cells; four such cells are produced in all.

The reproductive events in the life of an organism from zygote to zygote constitute its *life cycle* (fig. 9 and fig. 10). The development of sexual reproduction opened up great opportunities for exploitation of many different life cycles based on the alternation of haploid and diploid generations. In animals, the diploid state (the adult composed of 2N cells) has become dominant (fig. 10), with the haploid stage expressed only in the short-lived gametes.

Figure 8 (*opposite page*) Mitosis and meiosis shown diagrammatically. Explanations in text.

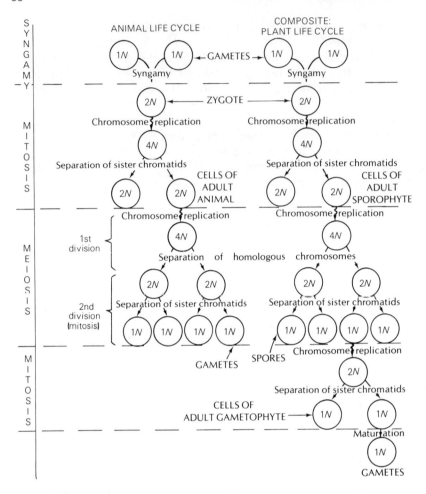

Figure 9 Animal and plant life cycles shown diagrammatically. Explanations in text.

In plants, the situation is more varied. As in animals, the zygote undergoes mitosis to form an adult plant (the *sporophyte*) composed of diploid cells and, from some of these cells, haploid cells (*spores*) form as a consequence of meiosis. The spores, unlike gametes, undergo mitosis to form an adult plant (the *gametophyte*) composed of haploid cells. Some of these cells become specialized as gametes; male and female gametes fuse to form a zygote, and the process begins anew.

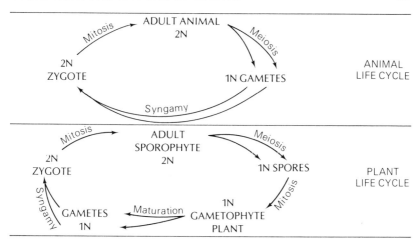

Figure 10 Animal and plant life cycles, diagrammed simply. Compare with figure 9.

Different plants emphasize different aspects of the life cycle. The green moss plant, for example, is the gametophyte (haploid) stage; the most primitive plants also have haploid-dominant cycles. Ferns and other higher plants are sporophytes (diploid). It is interesting that both metazoans and advanced metaphytes emphasize the diploid phase. The reason for this is that a double dose of chromosomes offers the organism more genetic options than does a single dose, and therefore provides more potential for adaptation in the physiological sense.

The evolutionary significance of sexual reproduction cannot be underestimated. There are two main advantages to the process:

1. The opportunity to exploit the possible variations of the haploid-diploid life cycle.
2. The fusion of genetic material from *different individuals* in the formation of the zygote. This is the process of genetic recombination, and it assures that each zygote is a unique genetic experiment. The result is that there is more genetic variability available in a population than there would be in an equivalent asexually reproducing population. The greater variability gives natural selection more material to work with, and this leads to more rapid rates of evolutionary change. The incredible morphological diversity of plants and animals on earth today and in the past is the product of the realized evolutionary potential of sexual reproduction.

Summary

We have looked very briefly at events marking the first 2.5 BY of life on earth. This brings us to the next 700–900 MY during which most of the types of organisms with which we are familiar came into existence. Evolution during the first part of the Precambrian appears to have been very slow, and this probably resulted from the fact that most of it took place among procaryotic organisms that perfected physiological adaptation mechanisms, and that reproduced only by binary fission with little recombination of genetic materials. Of these, the blue green algae reached an evolutionary dead end, or, looked at in another light, complete success; they developed organisms that were so physiologically adaptable that they ceased to change by natural selection.

The other procaryotes underwent a very extensive adaptive radiation, developing new metabolic types able to utilize the environment in many different ways. Both aerobic and anaerobic bacteria came into existence, some were chemosynthetic autotrophs and others were able to fix various elements and simple molecules from the atmosphere.

Eucaryote cells developed. Possibly the course followed the usual method of evolution by accumulation of small changes. The facts, however, suggest evolution by symbiosis — a radical departure from the more usual interpretation.

One of the major innovations during the Precambrian was the development of mitosis and, later, sexual reproduction by eucaryotic organisms. This development opened the way for still more adaptive radiations, this time in terms of life cycles, and provided increased variability to populations by means of genetic recombination in the fusion of gametes.

A number of interesting and basic evolutionary concepts, including some that are evident in other parts of the record as well, are encountered in the evolutionary history of the one-celled organisms. Among these are convergent evolution, adaptive radiation, attainment of new levels of organization, invasion of new ecological subzones, multiple, or polyphyletic, origins of new major groups of organisms, and the interaction of physiological and evolutionary types of adaptations. A knowledge of these processes paves the way for a close look at bigger, but not necessarily evolutionarily *better*, things — the Metazoa.

IMPORTANT CONCEPTS

AEROBES: organisms either able to metabolize and survive in the presence of oxygen or capable of utilizing oxygen in carrying out life processes.

ANAEROBES: organisms that must exist and carry out metabolism in the absence of oxygen.

ALGAE: a general designation for a wide variety of single- to multicellular plants lacking multicellular organ systems. **Procaryotic plants:** *Blue green algae*—lacking definite tissues; form unicellular and multicellular aggregates. **Eucaryotic plants:** *Green algae*—simple-, single-, and multiple-celled. *Brown* and *Red algae*—seaweeds, mostly multicellular with some differentiation of function in different parts of the body. *Yellow green* and *Golden brown algae*—mostly unicellular. All algae have chlorophyll, but there are different types in several of the groups.

AUTOTROPHS: organisms which synthesize their own nutrient materials from nonnutritive molecules.

BIOHERM: a reef made up of organically produced materials, usually from algae and corals.

CENTRIOLES: small cylindrical bodies formed from the splitting of a centrosome. During mitosis and meiosis in eucaryotic cells, centrioles migrate to opposite sides of the nucleus and give rise to the *spindle apparatus*.

CENTROSOME: a small extranuclear body that is thought to have evolved from the base of a flagellum in the symbiotic theory of the origin of eucaryotes. In mitosis and meiosis, it divides to form the centrioles.

CHEMOSYNTHESIS: the production of nutritive materials from carbon dioxide and water using energy from other chemical sources. (*Not* photosynthetic.)

CHROMOSOME: each of the several threadlike or rodlike bodies containing genetic material and other substances that stain darkly during cell replication. Chromosomes in procaryotes are single circular structures; in eucaryotes, they are numerous linear structures.

DIPLOID: a cell with two sets of chromosomes.

EUCARYOTE: a cell with a well-defined membrane-bound nucleus and various membrane-bound organelles (mitochondria, plastids, etc.).

FLAGELLUM: a long, whiplike appendage that serves as an organ of locomotion in some bacteria and protozoans. *Cilia* are similar, but are smaller and usually present in great numbers. A cross section of a flagellum, cilium, or centrosome shows a $9 + 2$ arrangement of protein fibrils.

FUNGI: multicellular plants lacking chlorophyll. They are all *parasitic* and *saprophytic* (living on dead organic matter) forms.

GAMETES: the mature haploid reproductive cells produced by meiosis.

GAMETOPHYTE: a stage in plant life cycles represented by an adult plant or tissue that is composed entirely of haploid cells.

HAPLOID: a cell with a single set of chromosomes.

HETEROTROPH: an organism incapable of synthesizing its own nutrient materials (proteins, carbohydrates, etc.). These nutrients must be taken preformed either from other living matter (autotrophs or other heterotrophs) or from abiogenically derived complex organic molecules such as those existing in the primordial organic *soup*.

HOMOLOGOUS CHROMOSOMES: found in diploid cells derived from the union of two haploid cells. Each chromosome from one haploid set has a structural and functional *twin* from the other haploid set with which it pairs and is separated during the first meiotic division.

LIFE CYCLE: the sequence of reproductive events in the lifetime of an organism from its inception as a zygote to its production of offspring or gametes.

MEIOSIS: replication of genetic material in a diploid cell followed by two successive cell divisions, resulting in a final product of four haploid cells. These haploid cells may be gametes (animals) or spores (most plants). The process characteristic of sexual reproduction.

METAPHYTA: multicellular, chlorophyll-carrying plants with cells forming distinct tissues and organs. Includes mainly vascular plants and mosses.

METAZOA: multicellular animals with cells forming distinct tissues and organs.

MITOCHONDRIA: organelles characteristic of eucaryotic cells. They are the sites of many of the crucial chemical reactions of life, particularly energy-releasing reactions. They are bound by membranes similar to cell membranes. In the symbiotic theory of the origin of eucaryotes, they are thought to have originated from bacterial symbionts.

MITOSIS: cell division in which chromosome replication is followed by division of the cell to produce two daughter cells, each with the same number of chromosomes as the parent. Technically, the separation of sister chromatids.

PHOTOSYNTHESIS: the use of radiant energy (light) to form complex organic molecules from carbon dioxide, water, and salts with the aid of a catalytic substance, usually chlorophyll.

PLASTIDS: organelles characteristic of photosynthesizing procaryotes. They contain chlorophyll and are the sites of many critical chemical reactions of life. In the symbiotic theory of the origin of eucaryotes, plastids are thought to have been derived from blue green algal symbionts. They, like mitochondria, are membrane-bound.

PROTISTA: a variously used term that includes organisms lacking definite tissues and multicellular organ systems. Includes everything except metazoa and metaphyta.

SISTER CHROMATIDS: the products of replication of a single chromosome. The two chromatids remain attached to one another until anaphase of mitosis or the second meiotic division, at which time they move apart on the spindle fibers toward opposite centrioles.

SPORE: in plants, the product of meiotic division. Undergoes mitotic division to form a gametophyte plant composed of haploid cells.

SPOROPHYTE: the spore-forming stage of the plant life cycle represented by an adult plant formed from mitotic division of a zygote and composed of diploid cells.

THALLOPHYTA: eucaryote algae and fungi. Plants which have little tissue differentiation into organs or organ systems.

ZYGOTE: diploid cell resulting from the fusion of male and female haploid gametes.

REFERENCES

Barghoorn, E.S. 1971. The oldest animals. *Scientific American* 224: 30–42.

Cloud, P.E., Jr. 1968. Pre-Metazoan evolution and the origins of the metazoa. In *Evolution and environment*. ed. E.T. Drake, p. 72. New Haven: Yale University Press.

———. 1974 Evolution of ecosystems. *American scientist* 62: 54–66.

Margulis, L. 1970. *Origin of eukaryotic cells*. New Haven: Yale University Press.

Schopf, J.W. 1970. Precambrian micro-organisms and evolutionary events prior to the origin of vascular plants. *Biological reviews* 45:341–47.

Schopf, J.W.; Haugh, B.N.; Molnar, R.E.; and Satterthwait, D.F. 1973. On the development of metaphytes and metazoans. *Journal of paleontology* 47: 1–9.

5

The metazoans

Perspectives

Most animals we encounter in day-to-day living are metazoans, and most of the plants are metaphytes. A large share of the interpretations of evolutionary processes, until the advent of molecular biology, have been based upon information from these two groups. Metazoans first appear in the fossil record in rocks formed about 650 MY ago. From the beginning of the Cambrian period, about 620 MY ago, to the present, they have a good record. The majority of interpretations of past history of life have been based on fossils from this relatively short span of time; short, that is, when viewed in the perspective of the full time of life on earth.

Some classifications recognize 25 or 26 phyla of metazoans, others as few as 10. Figure 11 shows some of the major groups and their phylogenetic relationships. Most of the major groups that have a good record appeared as fossils between 650 and 550 MY ago. Some phyla, mostly made up of soft-bodied, parasitic animals, have no record at all, and some have only a very sparse record. The major groups are listed, and some of them are illustrated at the end of this chapter.

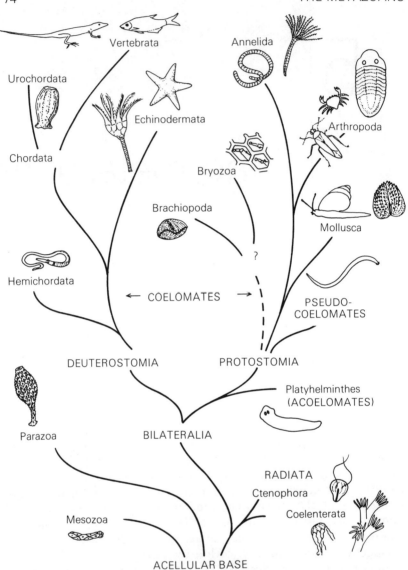

Figure 11 A general phylogeny of animals, showing the major groups and sketches of representatives. Note the major division into Radiata and Bilateralia and the latter into Deuterostomia and Protostomia. See the glossary of chapter 5 and plates I–IV for explanations and figures of major groups of animals.

ORIGIN OF THE METAZOA

Once they were well established, the metazoans left a fossil record which allows interpretations of their evolutionary history. The record of origin, however, is just the opposite. It poses many problems, most of which cannot be answered without considerable speculation. In this chapter, we will first look at some of these problems of origins and then at general patterns of radiation that took place once the major groups had been established. Attention is directed to animals rather than plants, although a similar study could be made for the latter as well. The principles are the same.

Origin of the Metazoa

The major problems here are easy to phrase and difficult to answer. Abruptly, at the beginning of the Cambrian period, rocks which carry abundant fossils of metazoans were deposited. Prior to that time, very few metazoan fossils occur, although the rocks in many formations seem to be suitable for their preservation. The principal exceptions are fossils of the Ediacaran fauna, about 1000 specimens in all (fig. 12). These have come from Australia and some other parts of the world and are probably about 650 MY old. If sponges (*Parazoa*) which are not strictly metazoans are included, six phyla are found in the lower Cambrian and four in the Ediacaran fauna. Some of the Cambrian phyla have two or more classes, showing a high level of diversity. Among these are such advanced animals as *echinoderms, molluscs,* and *arthropods*. By the end of the Cambrian and early Ordovician, all the phyla that have a good fossil record are represented, and occasional unusual finds show the presence of some of the other phyla as well.

It appears from this evidence as if there must have been a very long period of evolution prior to the Cambrian. If it is assumed that the first eucaryotes came into being about 900 MY ago, as the evidence suggests, then there is a period of nearly 300 MY in which this evolution could have taken place. The usual way of looking at this evolution is that it progressed gradually through a series of successive stages of increasingly advanced animals, as shown in figure 13A. Other possible patterns, with bifurcation of the two major groups or with multiple origins of several groups of metazoans, are shown in figures 13B and C. The big problem related to evolution as illustrated in figure 13A is that it would have taken a long period of time.

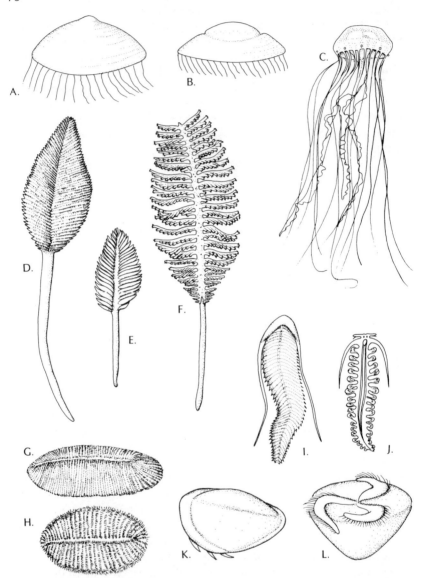

Figure 12 Animals from the Ediacaran fauna and some modern animals which they resemble. A and B, jellyfishlike animals; C, modern jellyfish; D and E, two genera of fossil sea pens, and F, a modern sea pen (a type of coelenterate coral); G, a fossil annelid, and H, a modern annelid worm; I, a fossil segmented worm and J, a modern segmented worm; K and L, animals of uncertain affinities. Redrawn and modified after figures in M.F. Glaessner (1961).

ORIGIN OF THE METAZOA

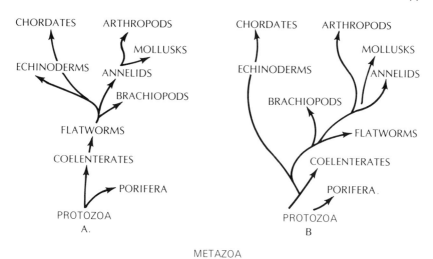

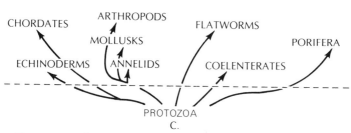

Figure 13 Three possible phylogenies of the major groups of animals. *A*, A serial phylogeny passing through successive stages of advancement, as represented by modern animals. *B*, A branching phylogeny in which two major groups are recognized as differentiating very early and going through the series of common steps. *C*, A polyphyletic phylogeny in which different major groups of metazoans arose independently from different groups of protozoans, or single-celled animals.

The 300 MY which probably was available is sufficient, but the record shows a curious thing. Bacteria, blue green algae, and eucaryotic algae are found in rocks deposited during this interval, but there is not a trace of metazoans. This poses a difficult problem, one we will look into but will not answer in a satisfactory way.

How good is the record?

The fossil record is notably incomplete. If, for example, we assume that there are about five million species living today and recognize that only about 200,000 are known for the full 3.5 BY of history, some idea of this incompleteness comes through. Throughout this long period of time, uncounted millions of species have come into being and have become extinct. In addition, the species that have been preserved come from those groups that fossilize readily, usually because of the possession of hard parts, and from environments where enclosing sediments formed and lasted until the present. The farther back we go into the fossil record, the less complete it becomes. Undoubtedly then, the record of life from the Precambrian is very incomplete and not a reliable basis for solutions of problems of origin of metazoans.

The full explanation of the absence of a record of early metazoans may lie in this fact alone. It has been suggested, for example, that the rocks in which fossils might have been preserved were eroded away, or that the places where metazoans lived were not those in which the rocks we see today were formed. On another line, it has been suggested that the early metazoans were all soft-bodied and were not preserved for this reason. All such suggestions may be in part or completely true.

Still, there are some facts that raise questions. First, of course, is the sudden appearance of fossils in rocks of Cambrian age, whereas similar rocks, formed earlier, do not have similar fossils. Many of these rocks, deposited between the time of origin of eucaryotes and the beginning of the Cambrian, as we have seen, carry well-preserved fossil remains of blue green algae, other types of algae, and bacteria. The environments in which these fossil deposits were formed appear to have been suitable for occupancy by metazoans. The known protistans would have provided a good food supply. If metazoans were present, even though they were soft bodied, they should have been preserved. The soft-bodied fossils that make up the Ediacaran assemblage are quite well preserved. If this sort of evidence is credited, then it would seem that the metazoans may have come into being rapidly, rather than over a long period of time. At least this is a good working hypothesis for additional investigation. It raises some other interesting questions.

Why and when did metazoans arise?

In summary, what seems fairly well established is that all evolution for the first 2.5 BY of life on earth was by procaryotes. Then, between 1300 and 900 MY ago, eucaryotes arose. Sometime after this, by 650 My ago, the first metazoans appeared.

It seems safe to assume that the eucaryote cells with the capacity for

ORIGIN OF THE METAZOA

sexual reproduction were a necessary step in development of metazoans. The biological basis for metazoans was thus established long before any representatives are found in the fossil record. The question then arises as to whether the apparent delay depended primarily upon biological factors or whether it resulted from a delay in the coincidence of necessary biological *and* physical factors. This is a much argued point.

The origin of metazoans, or for that matter of any group of organisms, requires a suitable environment. The question here is whether such an environment was present at the time the origin of metazoans first became biologically feasible. One aspect of the environment which has figured in the discussion of this point is the amount of oxygen in the atmosphere.

An oxygen-bearing atmosphere with much more than just a small percentage of oxygen was certainly necessary for the existence of metazoans. The earliest known forms were *pelagic* and lived near the surface of the sea. Oxygen was critical for their metabolism and also to provide an atmospheric shield from ultraviolet radiation by supplying an ozone shield. A principal source of atmospheric oxygen, according to many who have studied this problem, was the *phytoplankton* composed of microscopic, floating green protistans. These, too, needed protection from ultraviolet light. Such protection could have been provided by an atmosphere with about 1% oxygen, or about one-twentieth of what is present today. Until this level was reached and a further buildup of oxygen from phytoplankton had taken place, radiation of metazoans could not have occurred. This type of atmosphere, it has been argued, developed rather late in the Precambrian, thus delaying the appearance of metazoans until long after their biological basis had been established. The Ediacaran fauna presumably followed upon development of this atmosphere quite rapidly.

Such a physical limitation may have actually existed. But as always, with such slight evidence, some contrary facts do exist. For one thing, there is considerable evidence that oxygen-producing organisms were present long before this time, photosynthetic blue green algae as much as 2.7 BY ago. In addition, it seems probable that organisms were not the only source of atmospheric oxygen. Water vapor is dissociated by ultraviolet light into hydrogen and oxygen. How effective dissociation may have been as a source of oxygen depends upon how effective this process can remain as the oxygen content of the atmosphere increases. As soon as a strong ozone shield is present, then dissociation will virtually cease. Opinions vary widely and, consequently, estimates of the effectiveness of the inorganic source of oxygen differ. From what is known at present, the role of oxygen as the only limiting factor seems rather unlikely.

In this event, either some other limiting physical factors were active, or else the delay was strictly biological. If the latter is accepted, we must look into the origin of metazoans to try to see why and how they may have arisen and how, in view of this, a rather rapid deployment after origin might have occurred.

A plausible course of origin is as follows. Nonphotosynthesizing, free-moving, single eucaryotic cells under pelagic conditions had limited capacities for controlled motion. Except very locally, their movements were mediated largely by currents and eddies. Utilization of potential food resources was limited both by the restricted range and by the digestive potentials of the cells. Larger size, we may speculate, carried selective advantages of stability and controlled movement, with more effective procurement of a wide variety of foodstuffs.

Size may have been the selective advantage that was the key to the origin of metazoans. But larger size could be effective beyond very moderate limits only if more efficient means of locomotion developed. Improved mobility required body movement, augmenting and superceding the movement by cilia. Any changes which produced cells with musclelike contractile capacities would be of selective advantage. If these capacities occurred, they would tend to be preserved and transmitted to successive populations. This could have been the initial stage in the development of tissue differentiation leading to specialized systems with particular functions.

Beyond this, we may suppose, the development of other systems followed, each one taking on adaptive significance once size and mobility were attained. Coordination by nerve cells, sensing of the environment by rudimentary sense organs, and extracellular digestion of food might all arise. We know they did come to be. What is suggested is a way in which these characteristics might have arisen, as adaptive responses to an initial key change.

These adaptive processes might have gone on in several different lines, from different sources, for similar adaptive advantages would tend to follow increase in size in many primitive, multicellular, or multinucleate types. *Polyphyly* is a distinct possibility and, among other things, could have set the stage for rapid deployment of metazoans, as we will note a little later.

How much of this theory is really true? No one can say as yet. Given the known evolutionary processes and mechanisms, the knowledge we have of ancient life and environments, and the properties of

ORIGIN OF THE METAZOA

living remnants of ancient stocks, it is plausible. But critical stages are not known, and what we see as a logical series of events may, of course, be far wide of the mark.

From what did metazoans arise?

Beyond a concensus that metazoans came from eucaryote ancestors, there is little agreement. One hypothesis is that there was a single eucaryote source, that is, that metazoans are monophyletic. Following this assumption, the source has been variously sought among the following:

1. Unicellular protozoans (protistans); from flagellates or ciliates;
2. Multicellular plants;
3. Colonial, multicellular protozoans, various groups;
4. Multinucleate protistans, organisms with a single cell with many nuclei in common protoplasm.

All are plausible sources. That this statement is true and that some supporting evidence can be found for each possible source raises the possibility that there may have been more than one source of the metazoans, that is, they are polyphyletic. This idea has a distinct advantage from the standpoint of time, getting away from the requirement of a slow passage through successive levels of organization. The different basic organizational patterns of the several phyla could have been established separately in different lines rather than sequentially.

Cloud (1970), for example, has suggested that some of the basic metazoan types might have had the following sources:

1. *Parazoans*, the sponges, may have come from one-celled protistans, from a particular group known as *choanoflagellates*. In parazoans, there are flagellate cells with collars, similar to the collar cells in choanoflagellates.
2. The sea-comb-like *coelenterates* might have come from a *metaphytic* ancestry. Metaphytic, as used, refers merely to multicellularity and not to the development of organ systems. Roughly, the source of coelenterates is sought in algal fronds. This applies to *anthozoans* (corals and

corallike coelenterates) but does not specify a source for jellyfish.

3. Flatworms, or *platyhelminthes*, may have originated from multinucleated ciliates. In succession from very primitive, *acoelomates* (turbellarian flatworms lacking a coelom) may have come *annelids*. These seem to be not far from the sources of *molluscs* and *arthropods*. The source of *radiates* is also sought in the primitive flatworms. If flatworms were ancestral, then the radiates, radially symmetrical animals, arose from bilaterally symmetrical ancestors.

4. *Echinoderms* may have arisen from armored *dinoflagellates*. This suggests a completely separate origin of the *deuterostomus* echinoderms and *chordate* and *proterostomous* forms, the great majority of *invertebrates* with both anus and mouth present.

As before, these are merely suggestions of possible sources, each with some supporting evidence from moderns but almost nothing from the past, other than the existence of members of the Ediacaran fauna in the Precambrian. The source of *brachiopods* and other *lophophores* remains a puzzle.

To understand evolutionary events, modes, and rates of changes, it is necessary to have much more specific information on the actual course of events. Perhaps with new finds this may eventuate, but as yet many questions remain wide open.

Summary on origins

The problem of the apparent sudden appearance of metazoans has been the basis for many kinds of speculation, including those that require a special kind of evolution which involves catastrophically rapid modification. To propose such changes on the basis of the record, which is very incomplete, seems unwise in face of the fact that elsewhere in the fossil record, wherever it is at all complete, changes conform to a normal slow Darwinian kind of evolution.

On the basis of the best data available, the origin of metazoans appears to have taken place quite rapidly. The stage for rapid changes was set with the origin of sexual reproduction in eucaryotic cells. After a rather long period of time involving initial exploitation of this new development, several types of multicelled organisms at the lower threshold of metazoan organization appeared in the fossil record. The first evidence of this level is in the Ediacaran fauna from about 650 MY ago. Thereafter, over a period of about 30 MY, the basic phyla of the Metazoa appear to have been established. The implied rate of change is not implausible if it is assumed that there were multiple origins of the basic ground plans of the major subdivisions of the metazoans. At present, this pattern of origin seems to best explain the relatively few facts that are available. This is and must remain speculative until a much better record of the actual events is obtained.

Radiations of the Metazoa

The various phyla of metazoans do exist and several of them can be followed through the history of their evolution with little difficulty. Very generally, the unfolding of the potentials of the basic organizational structure of the phyla and classes they contain follows a threefold pattern:

1. Origin of the new level of organization, i.e., the phylum itself; classes, often in temporal sequence; and lesser units. The lowest level at which this phenomenon stands out is the family.
2. A period of development in which the innovations established at the new level are brought to maturity.
3. A phase of adaptive evolution in which the potentials of the organization are exploited to the fullest extent permitted by the environments in which the members of the group live.

An idealized diagram of this broad evolutionary course for a phylum is shown in figure 14. Figure 15–18 illustrate several different patterns exhibited by major groups. Many common features are seen—the rapid expansions (indicated by the balloons), subsequent rapid decreases in kinds, and, in many instances, persistence of a small branch of the radiation long after much of it has disappeared.

When the patterns are studied in detail and the organisms related to the physical and biological habitats, the reasons for the general pattern and the variations upon it appear. Expansions can be related to invasions of new environments, either occupied or unoccupied. In the second instance, replacement of the old by the new takes place.

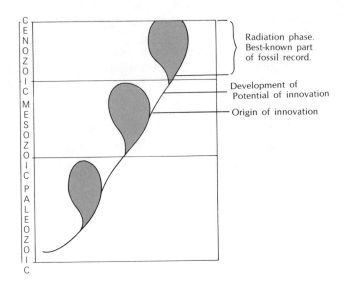

Figure 14 A general diagrammatic pattern of the course of radiation in most groups of plants and animals. Note the point of origin of an innovation, the maturation of innovations, and the successive adaptive radiation.

Restriction of the radiation in some instances is related to the replacement of the fully expanded group by another that is on the rise. In other instances, restriction follows marked alterations of the environments, eliminating many conditions to which the organisms are highly adjusted. The persistent remnants usually are types that have come to occupy some very stable environment which lasts through time. A muddy marine shore environment is of this sort. The various patterns differ somewhat, both because of the different radiative capacities of the organisms and because of the different environmental vicissitudes which they encounter.

Figure 15 shows several groups of brachiopods. First, it may be noted that these had their greatest success during the Paleozoic era and have been somewhat restricted ever since. Two major kinds exist, *inarticulates* and *articulates*. The absence or presence, respectively, of a hinge at the junction between the two shells is the distinguishing character. *Lingula* and *Terebratula* are living examples of the two types.

The *Lingula* type, which has been very persistent in time, inhabits muddy shores and is a burrower. It became adapted to this habit early

RADIATIONS OF THE METAZOA

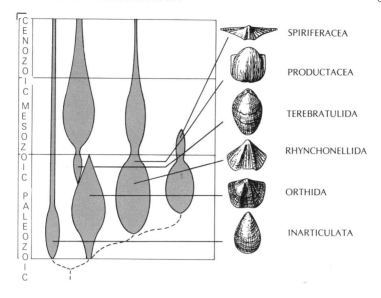

Figure 15 A diagram of the adaptive radiation of the brachiopods (see text for explanation). The envelopes of balloons roughly indicate the number of kinds present. The drawings to the right show the representatives of the various major groups.

and has changed but little. The articulate group inhabits a much wider variety of environments, and individuals usually live on the floor of the ocean rather than in burrows. As might be expected, the articulates have undergone much greater radiations. This has included adaptation to different modes of feeding and to different substrates. The ways of obtaining and utilizing food and of attachment to hard and soft sea floors are reflected in the structure and form of the shells. Each new type of brachiopod in the diagram developed innovations which made invasions of new environments possible. The increases and decreases of the different groups, indicated by the sizes of the balloons, reflect both structural modifications within the groups and changes of the environment during geological time.

With the beginning of the Mesozoic era, after a sharp depletion in unfavorable environments of the late Paleozoic, brachiopods increased somewhat, but they never attained the success they had in the Paleozoic. Largely, it seems, this was because of the development of an abundance of new competitors, especially among the molluscs.

Cephalopods, figure 16, illustrate a somewhat different set of patterns. These are highly organized molluscs with well-developed sys-

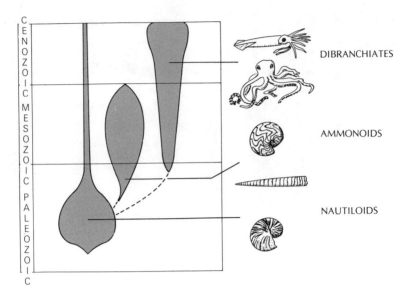

Figure 16 A diagram of the adaptive radiation of the major groups of cephalopod molluscs. See text for explanation.

tems. They have eyes which are similar in many respects to those of vertebrates. The earliest radiation involved forms with rather simple, external shells, the *nautiloids*. The only living representative is *Nautilus*. Some were pelagic, floating and swimming in open waters, and others inhabited the shore areas of the seas. During the middle Paleozoic, nautiloids gave rise to somewhat more complex cephalopods, the *ammonoids*. The latter evolved rather slowly after their origin and then underwent a major radiation during the Mesozoic. Their shell structure is very well known, but no representatives have lived to the present. In absence of intimate knowledge of the soft anatomy and physiology, the reasons for their great success are not very clear.

After being abundant and highly varied almost to the end of the Mesozoic, the ammonoids died out. This was at about the same time that the great marine reptiles became extinct and when the dinosaurs and other kinds of reptiles disappeared from the continents. Many explanations have been advanced to account for this extinction of widely varied groups of animals, and for the disappearances of the groups separately, but none is fully satisfactory. Surely the ultimate cause was ecological, but just what the major factors were is uncertain.

RADIATIONS OF THE METAZOA

During the late Paleozoic, another innovation took place, the development of cephalopods with eight tentacles, in contrast to the ten of the other groups. These new forms also had internal rather than external skeletons and were highly mobile. *Squids* and *octopuses* are living examples. This group had a modest radiation through the Mesozoic and has remained prominent in marine communities throughout the Mesozoic and Cenozoic.

The *echinoderms*, figure 17, illustrate still another variant of the common plan. Here, soon after the appearance of very primitive

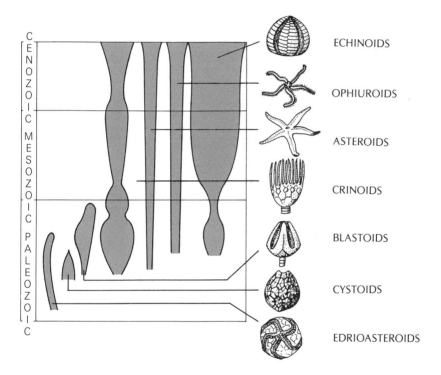

Figure 17 A diagram of the adaptive radiation of the major groups of echinoderms. A number of the early extinct groups have been omitted. See text for explanation.

forms, the *edrioasteroids*, a rapid divergence into a large number of classes took place. All in all, some 20 developed. The primitive forms were somewhat sedentary, encrusting rocks and shells of other animals. A five-fold, radial symmetry was superimposed upon the initial bilateral symmetry and persisted throughout the history of the echinoderms. One series of descendants, remaining largely fixed, developed stems which permitted the heads, or *calyces*, considerable motion. Successive innovations in the three-stemmed lines— *cystoids, blastoids,* and *crinoids*—resulted in changes in the complex water-circulating system, the water-vascular system, and in food-gathering structures. Two of these lines failed to survive the Paleozoic, but the crinoids underwent a series of pulsations of radiations and restrictions, first in the Paleozoic and then through the Mesozoic and Cenozoic eras. During the last, forms no longer fixed by stems and using their long arms, originally related to feeding, for locomotion became successful. This innovation appears to be one reason for their recent increase.

The other three lines illustrated for the echinoderms are free forms—*starfish* (*asteroids*), *brittle stars* (*ophiuroids*) and *sea urchins, sand dollars,* and similar animals (*echinoids*). Both starfish and brittle stars are inhabitants of the sea floor, from shallow waters to the abyssal deeps. While very successful in this broad habitat, they have not undergone very extensive radiations, as reflected in the narrow pathways in the figure. The environments which they have inhabited have persisted without marked changes, and no innovations carrying them beyond these general habitats have occurred.

Echinoids behaved differently. They underwent several pulses of extensive radiation and have developed a wide array of kinds in a variety of marine environments. Beginning with the Mesozoic, an extensive and complex radiation began and, with some minor restrictions, has continued to the present time.

These cases include types of patterns which occur with various modifications throughout the metazoans and are much the same as those in terrestrial plants. Figure 18 shows a similar pattern for the *tetrapods*, four-footed vertebrates. In many respects, the radiation pattern is similar to that of the cephalopods. In figure 19, a part of this is given in more detail to show some of the inner workings of the major "balloons." What is shown is typical for many phyla of metazoans. The record of the vertebrates includes animals familiar to most and, thus, is particularly good for illustration.

In figure 19, the numbers from 1 to 8 serve as reference points for discussion of the major events. The letters *A–D* specify more detailed events, but even these are well above the species level at which the basic evolutionary modifications are actually realized. In figures 20 and 21, skulls of some of the important kinds of animals in the sequence in figure 19 are illustrated. These events in figure 19 show some, but not all, of the significant modifications.

RADIATIONS OF THE METAZOA

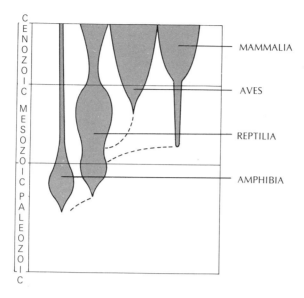

Figure 18 A diagram of the adaptive radiation of the vertebrates. See the text for explanation and the next three figures for elaboration.

1. The origin of reptiles from amphibians. Here, a major innovation occurred, the development of an egg which could mature on land, whereas that of amphibians requires an aquatic medium. This made possible exploitation of the opportunities of life on land.

2. From the very primitive reptiles, with very little change, came reptiles that gave rise to mammals. These were the *pelycosaurs*. Very little modification can be seen in the earliest members, and the radiation is essentially one with that of the very primitive reptiles. One group, A, became increasingly aquatic, feeding on fish and other freshwater organisms. B represents lines of herbivores, some partially aquatic, some terrestrial. C represents carnivores with several branching lines. One of these led to an innovation at 3.

3. This marks an innovation basic to the extensive radiation of *therapsids*, advanced mammallike reptiles. One of the major changes was an enlargement of the posterior parts of the brain, related to increased motor activity. It

may also be that there was an increase in *endothermy* at this stage.

4. This is the therapsid radiation, at the termination of which mammals came into existence. Members of the many lines penetrated a wide range of terrestrial habitats. *A* includes the lines in which there was rapid adaptive departure from the ancestral condition. Representa-

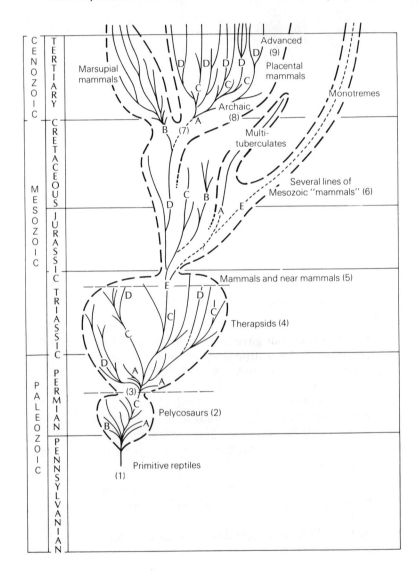

RADIATIONS OF THE METAZOA

tives became larger and more specialized but did not alter or mature the major innovations at 3. *D* represents several lines that became wholly or partially adapted to herbivorous diets. Note that this feature arose convergently in different lines. *C* designates several lines of carnivores, each distinctive in its adaptations, and each moving in the general adaptive direction of mammals, i.e., becoming more active, probably more endothermic, modifying the jaws and dentition, and locomotor apparatus. These lines represent extreme cases of parallelism, with members of separate stocks doing similar things but in very different ways.

One of these lines, possibly more than one, reached a threshold which may be called mammalian. This is at *E*, at or just prior to 5.

5. This is a time of transition from very mammallike reptiles to mammals. Some features of the animals existing at this level were reptilian and some were mammalian on the basis of our usual definitions. In other words, each was a mosaic of advanced and more primitive characters. Jaws, for example (figure 20), still retained the reptilian elements in the articulation. In fully developed mammals, these articular elements became ossicles of the middle ear. The brain was relatively larger than that in advanced therapsids, enlarged posteriorly and in the olfactory region and approaching the organization level of some of the most primitive brains of living mammals. All of these animals were very small and, in many, features of the skulls and dentitions are mammalian. At this time, many of the important features of mammals were initiated, but they were not consolidated into a full organization which lay basic to mammalian radiations.

Figure 19 (opposite page) A diagram of the course of evolution from primitive reptiles to mammals and the adaptive radiation of mammals. This is an expansion of part of figure 18, to show the details within the balloons. Each of the individual lines (or branches) represents a particular group of animals. These are indicated by the letters *A* through *E* in the various balloons. The numbers in parentheses, (1) through (8), indicate levels of development and times of shift from one level to the next. Figures 20 and 21 have sketches of the skulls of the various groups of animals keyed by the numbers and letters under the sketches.

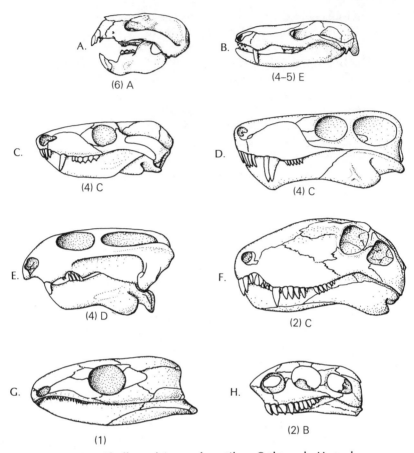

Figure 20 Skulls and jaws of reptiles, C through H, and very primitive mammals, A and B. The numbers in parentheses and the accompanying letters key these skulls to their positions in the diagram of figure 19, and to the text explanations. For example, (2)B is the B branch in the balloon of pelycosaur reptiles. B through H after E. Olson (1971), and A redrawn from A. Romer (1966).

6. This represents the modest radiation during the Jurassic and early Cretaceous periods of the lines established at 5. Several lines of small insectivorous forms—B, C, and D—developed somewhat in parallel. These are mostly known from jaws and teeth. Little evolutionary advancement took place; modifications were largely adaptive. Line A, however, is different, and includes somewhat larger herbivorous animals. These, the *multituberculates*, per-

RADIATIONS OF THE METAZOA

sisted into the early Tertiary long after the other lines had disappeared. *E* represents the living egg-laying mammals, the *monotremes*. Only two, the duckbilled platypus and spiny anteater, are known. Their source is uncertain but may have been among the multituberculates. There is no good fossil record.

7. At this level, true, or *therian*, mammals came into being. Presumably, there were innovations at this stage, but these are not well documented in the record. From what is known later, it may be assumed that the brain had made important advancements and that *endothermy* was well developed. These were true mammals in all respects. The opossum, a living marsupial (fig. 21), probably is somewhat like the early members of the stock.

8. This represents the vast radiations of the mammals proper, *placentals* and *marsupials*. Shortly after the origin of the therians, the placentals, line *A*, and the marsupials, line *B*, separated. Both radiated in similar ways with a great deal of convergence of late members. The entries of *C*, under placentals, represent an early radiation, characteristic of newly established groups as they very rapidly take advantage of new opportunities.

Note that this radiation was delayed for some time after the origin of true mammals, just as the true mammals arose well after the appearance of the first mammals. Two factors are involved in this phenomenon, which is common in the patterns of innovation and radiation. First, lag may be due to failure to attain full expression of the newly acquired level of organization. Second, the development of a suitable environment for radiations may not be coincident with the origin of the biological capabilities. In the case of placentals, for example, radiation did not begin until after the extinction of the major reptilian groups. This probably removed a barrier to their radiation. This may be a much oversimplified explanation, but, in any event, it does seem to be the case that the biological basis for radiation was established well before the radiation actually began.

The early radiations, *C*, produced many adaptive types, but the general body organization and, pre-

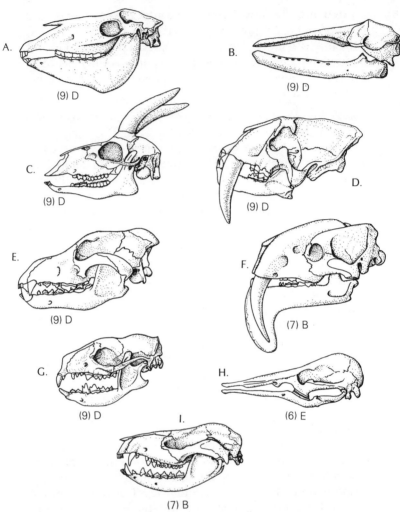

Figure 21 Skulls and jaws of various mammals, keyed to figure 19 by numbers in parentheses and by letters accompanying them and to explanations in text. A, C, D, E, F, H redrawn after A. Romer (1966), and D, G, and H mostly after W.K. Gregory (1951).

sumably, physiology was archaic as compared with that of most living mammals. From this basal radiation, many stocks were sorted out and these led to the later radiations of the Tertiary and to modern mammals, indicated by the letter D under placentals in the diagram. These

SUMMARY OF RADIATION

included *carnivores*—catlike forms (cats, civets, hyaenas) and doglike lineages (dogs, bears, raccoons, mustelids); many herbivores—*perissodactyls* (horses, tapirs, rhinoceroses), *artiodacyls* (pigs, camels, antelope, bison, sheep and goats), and *elephants*; *marine mammals* (seals, whales); *rodents* (rats, mice, guinea pigs); and *primates* (lemurs, monkeys, apes, man) to name a few. Some skull types are shown in figure 21. These placentals exploited a wide range of environments, many more than did the reptiles, utilizing the potentials of homoiothermy, brain structures, and reproductive capabilities.

Line *B*, marsupials, underwent a similar but less extensive radiation. Their reproductive processes appear to have imposed some limitations on the capacities to compete with placentals, and where similar products of the two lines have come into competition, the marsupials have fared badly. When, as in Australia, they developed without placental competition, they were very successful in occupying many different environments.

From the primates came man. In some respects, although his nonneural morphology is generally rather unspecialized, man, or *Homo sapiens,* represents a new level of organization in the development of his brain and the emergence of consciousness. Whether this will be a successful innovation or not is an open question. In the past, the great majority of innovations have failed, and the course of evolution followed the path of those which were successful. Man may fail in this respect, but if he does not, the innovation of consciousness and ability to control the environment may alter the course of evolution to the extent that many of the patterns we have studied in this chapter will no longer apply.

Summary of radiation

Once the metazoans appeared in fair numbers, radiations of each of the separate phyla took place rapidly. The records, however, give good evidence for the course of evolution of only a limited number of the total phyla. Those for which the record is good reveal a fairly constant pattern of change, with alternating times of innovation, bringing new levels of organization, and consequent adaptive radiations. Each major group has its own patterns, but within each the same sorts of events took place, occurring under different guises as determined on one hand by biological properties of the organisms and on the other by the environments in which they took place. This produces many variations on the common theme. Man probably represents the latest of such major innovations. The implications of his attainment of consciousness are immense, but the outcome of this evolutionary *experiment* are as yet unpredictable.

IMPORTANT GROUPS OF ORGANISMS

(see plates I–VI)

ACOELOMATES: metazoans lacking a body cavity. Platyhelminthes and Nemertina.

ANNELIDA: polychaetes, marine worms; earthworms, leeches. Segmented worms with systems and organs well developed. Marine, freshwater, and terrestrial. Fossil record to Precambrian.

ARTHROPODA: animals with jointed appendages. The important groups are: (1) Trilobita, extinct, three-lobed arthropods with an excellent Paleozoic record. Abundant in early Cambrian deposits; (2) Arachnida (spiders, ticks, mites, scorpions). Record to the Cambrian; (3) Insecta, the most numerous of the Metazoa. Record to the Devonian; (4) Crustacea (ostracods, crabs, crayfish, lobsters, shrimp). Record to late Cambrian. All arthropods are highly organized metazoans, segmented, with chitonous exoskeletons, and well-developed body systems. The various groups radiated into many different environments.

BILATERALIA: metazoans with symmetry such that a single vertical longitudinal plane divides them into two equal halves.

BRACHIOPODA: modern representatives called *lampshades*. Bivalves with dorsal and ventral valves. Feeding facilitated by ciliated fold around the mouth, the *lophophore*. Marine. Excellent Paleozoic fossil record.

BRYOZOA: sometimes called *moss animals*. Colonial, lophophore-bearing organisms. Many secrete calcareous skeletons. Good Paleozoic fossil record to the Ordovician.

CHAETOGNATHA: arrow worms, record in middle Cambrian, no other certain.

CHORDATA: Urochordata (tunicates and sea squirts); Cephalochordata (lancelets, *Amphioxus*); Vertebrata, backboned animals (fish, amphibians, reptiles, birds and mammals). All chordates have a dorsal longitudinal support structure, the *notochord*. Record back to the Ordovician for the vertebrates.

COELENTERATA: simple metazoans with a hollow gut. Jellyfish, hydras, sea anemones, corals.

DEUTEROSTOMIA: metazoans in which the mouth is a secondary opening, formed during *ontogeny*. Near the original opening, the anus forms. The cells are indeterminate through the early phases of cell division.

ECHINODERMATA: (starfish, sea lilies, sand dollars, sea urchins). Adults radially symmetrical with a basic five-fold symmetry. Characterized by a water-vascular system used in locomotion and feeding. Fossil record to early Cambrian.

ECHIURIDA: small group of genera with segmented larvae. No fossil record.

HEMICHORDATA: "Protochordates" (acorn worms, pterobranchs, possibly the extinct graptolites). If graptolites are included, the fossil record goes back to the Cambrian. Pterobranchs are known from the Ordovician.

MOLLUSCA: the important classes are: (1) Monoplacophora, very primitive with some segmentation (most molluscs are unsegmented). Ancient fossil record, possibly Precambrian; (2) Pelecypoda (clams). Fossil record to early Cambrian; (3) Gastropoda (snails). Fossil record to early Cambrian; (4) Cephalopoda (squids, octopuses, nautilus). Fossil record to late Cambrian. All molluscs have moderately well-developed body systems. Most produce an external skeleton formed by a mantle.

ONYCHOPHORA: somewhat intermediate between annelids and arthropods and rather caterpillarlike in appearance. *Peripatus,* living example. A representative in middle Cambrian, but no other record.

PARAZOA: the sponges (Porifera, or pore-bearing animals). They have a simple cellular organization without well-defined systems. Skeletons are composed of spicules of calcium carbonate or silica and/or protein fibers. A fair fossil record to early Cambrian.

PHORONIDA: about 15 species of marine, tube-forming, wormlike animals. No fossil record.

POGONOPHORA: the bearded worms. A few species of marine worms. No fossil record.

PROTOSTOMIA: a major subdivision of Metazoa. The original opening to the gut in ontogeny develops into the mouth in the adult. The anus, if present, is a new or secondary opening. The cells are *determinate,* that is, the developmental fate of each cell is established during initial cell divisions in the embryo.

PROTOZOA: unicelled animals (Protista or Animalia). Many different groups may be included, some classified as plants or animals as suits the interpretation of particular specialists. (1) Flagellates, having flagella as locomotor structures. Some photosynthetic organisms belong here, plants? (2) Choanoflagellates, protozoans whose cell is a collar-cell, very like that found in the multicellular sponges (Parazoa). (3) Dinoflagellates, some photosynthetic, many not. Plants or animals? Some colonial. (4) Ciliates, having many cilia (flagella) as locomotor apparatus. Mostly unicellular, but some colonial and some multinucleate.

PSEUDOCOELOMATES: animals with a *false* body cavity, developed differently from that in all other animals with a body cavity—the Coelomata. Includes Aschelminthes, Acanthocephala, Entoprocta.

RADIATA: radially symmetrical animals. (1) Coelenterata (hydras, corals, the Anthozoa; jellyfish, sea anemones). Simple organization, but with some well-developed systems. Many are colonial, with calcareous skeletons.

Mostly marine, some reef-forming. Fossil record to early Cambrian and perhaps Precambrian. (2) Ctenophora, sea walnuts, comb jellies. Marine and pelagic. No fossil record.

SIPUNCULOIDEA: small group of unsegmented marine worms. One record from the middle Cambrian. Others sporadic and uncertain.

TARDIGRADA: a small group of minute worms with characteristics more or less intermediate between annelids and arthropods. No fossil record.

WORMLIKE PHYLA: this is a catchall used for convenience. Many different and little-related phyla are covered by the term. (1) Platyhelminthes (flukes, flatworms, tapeworms). Very primitive Bilateralia, thought by some to lie near the base of this subdivision. No fossil record. (2) Nemertina, proboscis worms, no fossil record. Also nematode worms, ubiquitous in habit, many parasitic. No fossil record. (3) Acanthocephala, spiny-headed worms, intestinal parasites of vertebrates. No fossil record. (4) Entoprocta, small, sessile, colonial, freshwater, and marine animals. No fossil record.

REFERENCES

Beerbower, J.R. 1969. *Search for the past.* 2d ed. Englewood Cliffs, N.J.: Prentice-Hall, Inc. chapters 9–21.

Cloud, P.E., Jr. 1969. Premetazoan evolution and the origin of Metazoa. In *Evolution and environment.* ed. E.T. Drake, New Haven: Yale University Press. p. 72.

Glaessner, M.F. 1961. Precambrian animals. *Scientific American:* 72–78.

Gregory, W.K. 1951. *Evolution emerging.* vol. 2. New York: Macmillan Co.

Olson, E.C. 1965. *The evolution of life.* New York: New American Library Inc. chapters 6–10.

———. 1971. *Vertebrate paleozoology.* New York: John Wiley and Sons, Inc.

Romer, A.S. 1966. *Vertebrate paleontology.* Chicago: University of Chicago Press.

PLATE CAPTIONS

PLATE I.   Various primitive and intermediate level invertebrate Metazoa. A, A jellyfish or medusa, *Velella lata*. A coelenterate. B, A flatworm, *Gyrocotyle fimbriata*. An aceolous, primitive metazoan. C, A sea anemone, *Adamsia* sp., a coelenterate. D, An inarticulate brachiopod of the linguloid type. Similar to some of the very early brachiopods. *Glottidia pupanidatum*. E, A leech, one of the unsegmented annelid worms. F, The garden worm or angle worm, a segmented annelid.

PLATE II.   Various representatives of the Arthropoda. A, The white-veined or morning sphinx moth, a lepidopteran. B, The Florida spiny lobster, a crustacean. C, A grasshopper, an orthopteran insect. D, A sheep crab, one of the crustaceans. E, The praying mantis, another orthopteran insect. F, Coltalpa beetles, male and female, on a thistle. A coleopteran insect.

PLATE III.   Two molluscs, A and B, and four echinoderms, C through F. A, A chiton mollusc in dorsal view, related to gastropods. B, An extinct ammonite cephalopod from the Cretaceous period. Note the complex sutures where the septae join the outer shell, which is missing. C, A brittle star, an ophiuroid echinoderm. Note 5-fold symmetry, but with one arm, pointing to top of page defining a bilateral aspect as well. D, A different type of brittle star, with very large central disc. E, The common sea urchin, seen from dorsal side. F, the common starfish, *Asterias forbesi*. Note as in C and D, the 5-fold symmetry, but with some evidence of bilaterality.

PLATE IV.   Four vertebrates, A through D, and two molluscs, E and F. A, A mammal, the common coyote, *Canis latrans*. B, A fairly generalized teleost fish, the perch. C, An amphibian, the red-legged frog, *Rana aurora*. D, A reptile, the pearl lizard, *Lacerta lepida*. E, The garden snail or apple snail, a terrestrial mollusc, a gastropod. F, The common Atlantic octopus, *Octopus vulgaris*, an unshelled cephalopod.

PLATE V.   A and B, two subspecies of the butterfly *Colias meadii*, one of the sulfur butterflies. These show melanism (or darkness) related to the temperatures in which they live. A, This subspecies lives above the tree line in Colorado at an altitude of over 10,600 feet. Note the light markings around the edge of the wings and the somewhat lighter ground color. This pattern is even more pronounced on the underwings and also particularly evident in the coloring, which is a yellow brown. B, This subspecies lives below 6,000 feet in warmer temperatures. Note the dark wing margins and darker background, also emphasized in natural colors. These are both females. See text, chapter 7, for discussion. C and D are two butterflies—the Monarch, *Danaus plexippus*; and the Viceroy, *Limenitis archippus*. The latter is a Batesian mimic of the former. See discussion in text, chapter 7. The two are not closely related as shown by the differences in the larvae, E of the Monarch and F of the Viceroy.

PLATE VI. Algal fossils from Precambrian and comparative specimens from modern floras. A through D, unicellular microfossils from cherts of the Precambrian. A, Fig Tree group, early Precambrian; B, Gunflint iron formation, middle Precambrian; Skillogalee dolomite, late Precambrian; D, Bitter Springs formation, late Precambrian; E, Unicellular microfossils from the Clarno chert, early Tertiary (Eocene); F, Decussate quartet of algal cells from the late Precambrian; G, colonial algae from the Paradise Creek formation to be compared with H, Modern *Eucaspis* colonial forms; I, Filamentous algae from the Bitter Springs formation, late Precambrian, to be compared with J, modern algae of the family Nostocaceae (see N below). K, L, M, Notocacean filamentous algae from the middle and late Precambrian. Note L and M from the middle Precambrian have cystlike structures, enlarged heterocyst cells like those found in moderns, at H. N, Drawings of structure of different families of blue green algae based on living representatives. For comparison with fossil forms. All after J. William Schopf, *Paleobiology of the Precambrian: the age of blue green algae*. Published with permission of the author.

Credits for photographs and specimens used in illustrations in Plates I through V. We wish to express our appreciation to the following institutions and persons for permission to use the specimens and kodachrome slides from which the figures were prepared.

The Herbarium of the University of California at Los Angeles, under the direction of Professor Mildred Mathias. The photographs used were by Mary V. Hood and A.W. and Mary V. Hood. Plate I, figures A, E, F; Plate II, figures A, D, E, F; Plate III, figure D; Plate IV, figures A, C, D, E.

Marineland of Florida. Plate I, figure C, photograph by Nat Fain. Plate II, figure B; Plate III, figures E, F; Plate IV, figure F.

Museum of Natural History, Los Angeles County, California. Section of Entomology, by courtesy of Dr. Charles M. Hogue. Plate II, C; Plate V, all figures. Figures C, D, E, F, from slides; C and D, photographs by R. Pence; and F, based on a painting by C.M. Dammers. Figures A and B from photographs of original material by K. Bezy. Section of Ichthyology. Plate IV, figure B, by courtesy of Dr. Robert J. Lavenberg.

Field Museum of Natural History, Chicago, Illinois. Courtesy of Dr. Matthew H. Nitecki. Plate III, figure B.

Eastern Illinois University, Charleston, Illinois. Dr. Frederick Schram. Plate I, figure B.

University of Southern Florida, Tampa, Florida. Dr. Joseph K. Simon, department of Zoology. Plate I, figure D; Plate III, figures A and C.

A.

B.

C.

D.

E.

F.

A.

B.

C.

D.

E.

F.

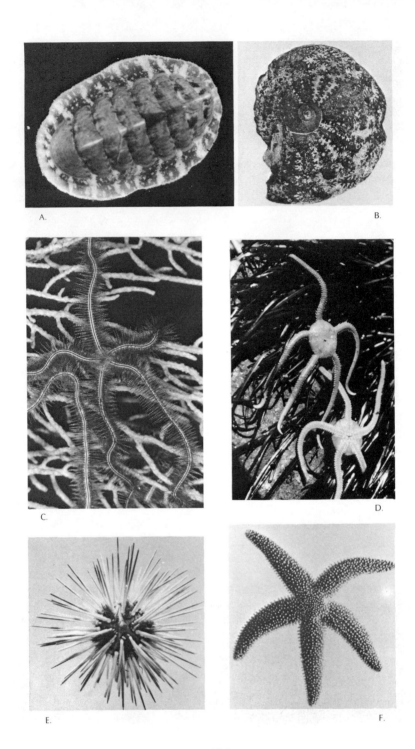

A.

B.

C.

D.

E.

F.

A.

B.

C.

D.

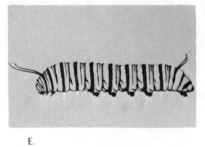

E.

F.

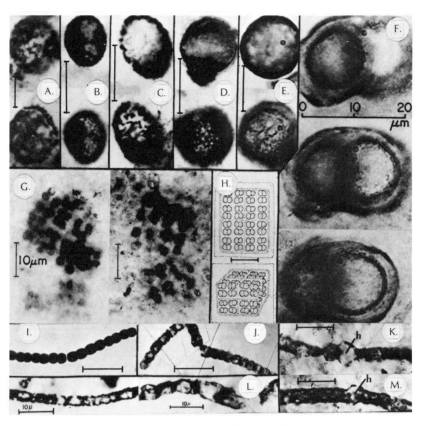

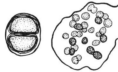

Chroococcaceae

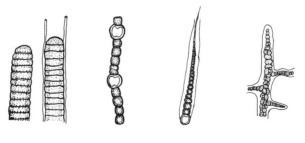

Oscillatoriaceae   Nostocaceae   Rivulariaceae   Stigonemataceae

6

Species, variation, and evolution

Perspectives

It will be instructive to turn back for a moment to classical Darwinian evolution to see how more recent studies have reshaped its basic concepts. Biological change, as Darwin saw it, resulted in the origins of new species; the process was directed by natural selection which differentially favored the survival of certain variants in a species as they engaged in a continuing "struggle for existence." For natural selection, he was able to amass an impressive body of data. However, as Darwin himself admitted on more than one occasion, there was no known mechanism for the origin of the variants. This, in the eyes of supporters and critics alike, constituted a serious shortcoming.

It is unfortunate for evolutionary theory that Darwin and his contemporaries were unaware of the genetic experiments of Gregor Mendel, a Czechoslovakian monk, which were actually completed before publication of *The origin of species*. As it was, some forty years elapsed before Mendel's work was brought to light. When it finally was rediscovered, however, there was little doubt that Mendel's principle of particulate inheritance was the missing piece of the evolutionary puzzle. Subsequent genetic research has elucidated the nature of the particles, or *genes,* and given us a greal deal of insight into the mechanism of inheritance both with regard to the transmission of variations and to the fidelity of replication.

Still other additions and refinements have been suggested to complement Darwin's ideas, and some of these have been incorporated into modern theory. For example, it is not the species, but the population which is now considered the *unit of evolution.* Ecological and behavioral studies have enriched our understanding of the selective process, and biochemistry is continually providing new insights into the unbelievably complex workings of the genetic materials themselves. It is a tribute to Darwin that none of these discoveries has done serious violence to his basic tenets, especially those related to selection.

The totality of these innovations as they have been superimposed on pure Darwinian theory is variously called the *synthetic,* or *Neodarwinian,* theory, and it is one of the most fully validated generalizations in all of science. It gives us a mechanical explanation of what we earlier termed the "most general statement of organic evolution" (see page 7); that is, that all forms of life arose by gradual changes from an original life stock.

In this chapter, attention is focused upon the nature of species, the source and inheritance of variation, and the role of natural selection in guiding the course of evolution. The interplay of the random and undirected processes which give rise to variation, and the channeling of these in particular directions appropriate to biological and physical characteristics of particular times and places, forms the basic theme. Evolution so conceived is a dialectical process in which activities of two very different realms combine to produce a coherent end result—the flow of evolution through time.

The nature of species

The term *species* in its most general sense means a distinct sort or kind of something. However, it has come to be most closely associated with biology, and in that context the word has had many usages. Before the eighteenth century, species represented a rather poorly defined assemblage of organisms that were alike in one or several features; as such, it was mainly a convenient way of making associations for one purpose or another. With the advent of biological classification in the Linnaean binomial framework, the species became the basic unit of classification. It included all organisms of an assemblage that resembled each other more than they resembled

THE NATURE OF SPECIES

other organisms; that is, it designated a particular kind of plant or animal. Variations exist among members of species so defined, but they are not sufficient to override the similarities that distinguish its members from those of other species.

Evolution may be considered to have taken place through the formation of new species in a process much like that envisioned by Darwin. This may happen in two ways, as diagrammed in figure 22. Either a parental species splits into two or more derived groups, or the parental species becomes gradually transformed into a new species as a result of gradual change with time. In both instances, there is some sort of *isolation* of derived groups from each other and from the parent group; isolation in the first case may be *spatial* or *behavioral*, and in the second case it is *temporal*.

Isolation in the case of splitting almost always has a geographical, or at least a spatial, component. To effect speciation it must render interbreeding between the isolated groups impossible or highly unlikely. Any one of a wide range of natural events can result in such isolation. On land, for example, the changing of a course of a stream, changing landforms, changing climates, and major fires are effective. Portions of widely ranging species which occupied both highlands and valleys when climates were cool may be isolated on separate uplands when temperatures in the valleys rise as a result of a general increase in temperature. In lakes and seas, similar changes in physical parameters—salinity, temperature, or physiography—are equally effective as isolating agents. Species arising from complete separation of this sort are said to be the product of *allopatric* speciation. Most species probably are formed in this manner (fig. 22A).

A more subtle form of isolation involves habitat partitioning, in which different groups undergo behavioral shifts away from behavior patterns in the parental species. The behavior shift may involve food preferences, breeding times, recognition signals, or any number of different behaviors affecting reproduction. The isolation may be *temporal* (groups breeding at different times of the year, as in many plants), *spatial* (groups feeding on different resources within the same habitat), or *recognitional* (closely related birds or insects utilizing slightly different calling songs). The isolation is often not as complete as it is in splitting, and speciation may take considerably more time if, indeed, it occurs at all. This type of isolation results in *sympatric* speciation, and has been the focus of much controversy in recent years.

Once isolated, by whatever means, the separate parts of the original species will tend to become modified in different ways, since it is virtually impossible that their suites of genetic variability and the

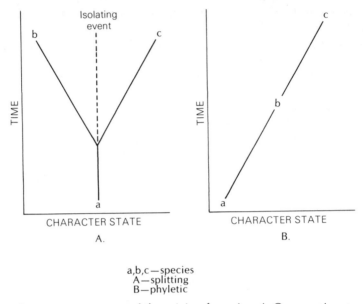

Figure 22 Two ways of the origin of species. A, One species, a, splits to form two, b and c, by divergence. Isolation is generally geographic. B, A gradual change with time of successive populations leads to forms sufficiently distinct to be called separate species.

selection pressures imposed upon them will be identical. As time progresses, the two groups will become increasingly different; that is, they will *diverge* behaviorally, morphologically, and genetically. The point at which the diverging groups become different species is, of course, difficult to specify, but in sexually reproducing groups, a commonly used (and by no means reliable) landmark is the inability of these groups to interbreed and produce fertile offspring. Likewise, the line between allopatric and sympatric speciation is not always easy to draw, and there are some authorities who seriously question the existence of speciation in a sympatric framework.

Nevertheless, if a definition of species based on discrete kinds is adopted, then evolution is the development of new groups (kinds) whose differences are of discernably greater magnitude than those between variants within a single species would be. There is nothing seriously wrong with such a definition; in fact, one very much like it has been the basis of much biological thought for many years.

Species are most often recognized and diagnosed on the basis of phenotypic characters, even now that other definitions of species are

THE NATURE OF SPECIES

in common use. A species is thus defined as a natural population whose members are distinct morphologically, physiologically, or behaviorally from those of all other such populations. The criterion of morphology is still the only biological basis available for determination of species in fossil organisms; it undoubtedly will continue to be used in paleontology and probably will continue as a mainstay for recognition of species among living organisms as well. It does, however, pose some serious problems.

Many of the problems are of a practical nature. For example, phenotypic characters may not always serve to differentiate between species, particularly in the fossil record. Many modern birds, lizards, and amphibians which are distinguishable on the basis of color and behavior patterns are almost impossible to tell apart on the basis of skeleton alone. Also, when individual characters or suites of characters are used, intergrades are often seen to occur. Such situations arise along geographic or temporal *clines* where the end members are distinct, but are connected by a continuum of organisms which provides an essentially complete morphological gradient between them (fig. 23).

A deeper problem arises from the fact that phenotypic characters may not clearly reflect the extent and nature of genetic difference, even though they are genetically determined. Males and females of a given species, for example, may differ in many morphological features in addition to those directly related to sexual structures—witness the mane of the male lion, the bright coloration of many male birds, or the enlarged claw on the male fiddler crab. The totality of genetic differences involved in all of these cases is trivial. All morphological features are removed from their genetic base (in the zygote) by a process of ontogenetic development, during which environmental influence may play an important role in determining the end-product.

To avoid such difficulties, an obvious approach is to use a strictly genetic definition of species which involves direct comparison of genotypes. Unfortunately, such information is tremendously difficult to obtain and interpret; and while it may be agreed that the ultimate test of degree of similarity should be through the comparison of the genetic constitutions of populations, the only available data concern individual *alleles*. A more direct *mass* approach to similarity is to test the degree of DNA hybridization (compatibility) between populations. The full potential of this method is yet to be explored, but so far it has not been particularly productive in dealing with detailed taxonomic problems.

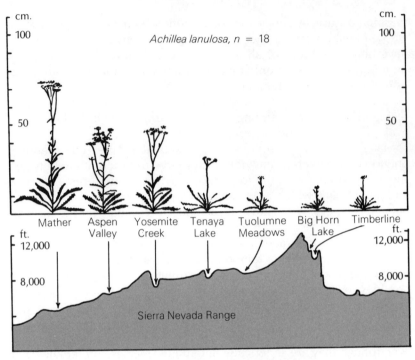

Figure 23 A series of populations of plants from a traverse across the Sierra Nevada range. The partially isolated populations are separated by distance and occur at different elevations. The clinal effect along the traverse is indicated by the different mean heights, as shown in the upper diagram, and by the localities and elevations in the lower diagram. The plant is *Achillea*, and the different populations represent a series of races. After J. Clausen, D. Keck, and W. Hiesey (1948).

For the most part, genetic similarity must be approached indirectly, leaving a gap similar to, but not necessarily as large as, that between phenotypic characters and genetic composition. Proteins such as albumin, cytochrome-c, haemoglobin, and histone are derived from genetic materials after relatively short sequences of biochemical events and thus may be considered more accurate indicators of relationship than gross morphology or behavior. Still other chemical processes utilizing different classes of organic compounds are being explored at the present time. These approaches to species determination and differentiation, along with studies at infraspecific levels, are becoming more and more important as skills and facilities for their

THE NATURE OF SPECIES

studies increase, and, significantly, they are becoming increasingly successful.

The idea of species as basic kinds of organisms founded on tangible characters, genotypic or phenotypic, will doubtless continue to play an important part in our thinking about the living world, simply because the species concept lies at the heart of our system of classification. In this context, a typological approach is useful and even necessary to insure meaningful communication among biologists and taxonomists. However, as our knowledge of variation, selection, and the dynamic relationships of organisms in their ecological settings becomes more refined, it becomes increasingly obvious that the unit of classification need not necessarily be the unit of evolution. When detailed studies of the processes and mechanisms of evolution are undertaken, attention is focused most strongly upon *populations*; indeed, a new discipline, population biology, has developed, and has contributed much to the ongoing Neodarwinian synthesis.

The reasons for choosing the population as the unit of evolution are not difficult to fathom. After all, it is populations which become isolated from one another and evolve into new species, and it is within populations that new genetic combinations and mutations are established. This is because species, generally speaking, are composed of several local populations which are only peripherally in contact as far as interbreeding is concerned. The result is that gene flow within the species is not random, and that differences can and do develop among populations, both in the genotypic composition and in the phenotypic expression of these differences. In the context of such a point of view, a species may be defined as a group of interbreeding natural populations that are reproductively isolated from other such groups (Mayr 1970); note, however, that this definition is inappropriate for asexually reproducing organisms.

Mayr's dynamic species offers a useful way to look at modern populations in flux, but when it is placed in the framework of geologic time, it loses much of its power. The fossil record offers little evidence of the breeding preferences of extinct organisms; further, it obscures the boundaries, both spatial and temporal, between whatever natural populations actually were in existence. Consequently, the paleontologist, working as he does with phylogenetic approximations, tends to view a species as a four-dimensional entity. G.G. Simpson has defined an evolutionary species as a *lineage* (an ancestral-descendant sequence of populations) evolving separately from others and with its own evolutionary role and tendencies (Simpson 1970). Although Simpson's and Mayr's definitions appear to be very different, the main distinction between them is one of resolution; both stress isolation and interrelationships of populations, but each is constructed to interpret a different body of data.

It seems, then, that the definition of species depends heavily on the use to which it is put in a given instance. This situation should not be bothersome, since definitions are simply conceptual tools facilitating communication among workers who are studying common phenomena; and a phenomenon manifesting itself in a number of ways can be looked at profitably from numerous conceptual vantage points. For our own purposes, however, we shall adopt the definition which sees a species as a *reproductively isolated group of natural populations*. Evolution then becomes defined in a way we introduced much earlier, as a *change in the genetic composition of populations* (Dobzhansky 1951) or as the *change in proportions of genotypes in successive populations*. The way in which these genotypic changes occur and are maintained is the subject of the rest of this chapter.

Variation and natural selection

General

Natural selection is the differential action of environmental factors upon the variability expressed by individuals within a population. Variability has its ultimate source in the genetic materials, but it is expressed in most cases through phenotypes, upon which selection acts more or less directly. The unit of selection is the individual organism, whose success, in an evolutionary sense, depends upon its capacity to survive long enough to produce viable offspring. Evolution occurs when selective removal of certain individuals and reproductive success of others causes a change in the genetic composition of the population as a whole.

As we have noted elsewhere, the totality of characters in an individual results from interactions involving the whole spectrum of genetic characteristics of the individual plus effects produced by the environment within which the individual develops and lives. It is impossible, or at least impractical, to study all the characters in the phenotype of an organism, or even to separate the effects of genes from the effects of environmental influence. For this reason, most studies are concerned only with one or a few characters whose selective value can be assessed and whose genetic base can be approximated. What we are concerned with in the following section is the nature of this genetic base—its structure, properties, and capacity for generating variability.

Biochemical variability

Ultimately, genetic variability arises through changes, or *mutations*, in the information-carrying macromolecules called DNA. This remarkable substance is capable of replicating itself either totally or in part. When the entire double helix replicates, new DNA results. RNA copies short sections of single DNA strands, and it is the RNA, carrying instructions chemically specified by DNA, which is responsible for the synthesis of proteins. Proteins are used both as structural building-blocks and as enzymes which mediate reactions in all metabolic processes; indeed, certain enzymes catalyze the replication of DNA itself.

Since proteins are almost singularly responsible for the phenotype of an organism, it is not difficult to see how a change in the information coded in the DNA can have repercussions at all levels of organization. Most mutations are deleterious to the organisms bearing them, since vital functions are disrupted when certain enzymes are impaired. However, mutations which are not harmful stand a good chance of eventually being incorporated into the genetic structure of a population, and thereby can enhance its variability. The occurrence of mutations is essentially random; and chance, it would seem, has much to contribute to the outwardly orderly, directed process of organic evolution.

In order to understand how mutations produce their effects, we must briefly examine the chemical structure of the informational macromolecules. DNA is formed by two *backbone* chains (fig. 24), the axial components of the famous *double helix*. Each is composed of a series of *nucleotides;* that is, it is composed of alternating units

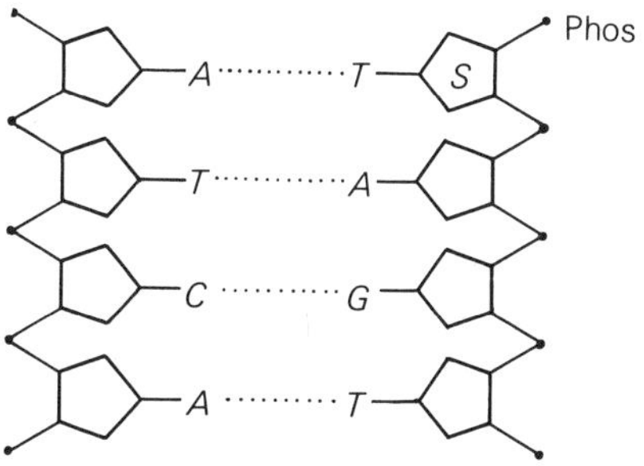

Figure 24 The basic *backbone* structure of DNA, showing the sugar (*S*), the phosphate (*Phos*) and the bases—*A*, adenine; *T*, thyamine; *C*, cytosine; and *G*, guanine. Also, see *nucleotide* in Important Concepts of chapter 3. The dotted lines are hydrogen bonds. See the text for explanation.

consisting of a phosphate and a sugar, to which is attached a *base* (see *nucleotide* in the Important Concepts of chapter 3). The bases of one chain can form hydrogen bonds with those of the other, and it is these bonds which tie the DNA superstructure together. There are four different bases:

> Thymine—*T*
>
> Adenine—*A*
>
> Cytosine—*C*
>
> Guanine—*G*

The chemical properties of the bases are such that *A* always pairs with *T* and *C* with *G*; taken together, these are the *letters* of the genetic code. Mutations at this level include additions or deletions of nucleotides or substitution of one nucleotide with its characteristic base by another.

The first step involved in the passing of information from DNA to protein is called *transcription*, and it involves copying a short segment of one chain into a molecule of RNA (fig. 25). RNA is similar in many respects to its parent molecule, but its sugar is ribose, not deoxyribose, it incorporates the base Uracil (*U*) instead of Thymine (which also pairs with Adenine), and it is single-stranded. There are three functional kinds of RNA—one carries the *message* specifying the amino acid sequence for a given protein and is called *messenger RNA* (mRNA); another binds to specific amino acids, there being one *transfer RNA* (tRNA) for each of the twenty or so amino acids in the cell; the last, rRNA, is involved with protein synthesis, and is part of a large globular molecule called a *ribosome*. Each of these is synthesized on DNA, and each can carry misinformation as a result of mutation of that substance.

The second step in the synthesis of protein is called *translation*, in which the message spelled out in base sequences on mRNA is chemically *read* and is translated into an amino acid sequence by a complex interaction involving tRNA, rRNA, and other substances. However, bases are not read individually, but are taken in groups of three; these triplets are called *codons* and constitute the words of the genetic code. A ribosome binding to one end of an mRNA initiates translation. The particular type of tRNA having the three necessary bases to bind to the initial codon (*GUG* pairs with *CAC*, etc.) will do so, bringing its distinctive amino acid along with it. The ribosome then moves along to the next codon and the process is repeated, with

VARIATION AND NATURAL SELECTION

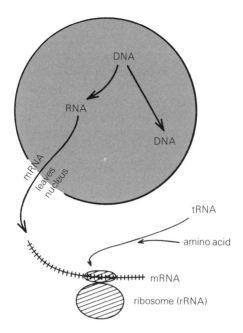

Figure 25 A diagram of the nuclear and extranuclear activities of the cell in the production of a protein enzyme. See the text for explanation.

the second amino acid becoming chemically bonded to the first. As soon as this happens, the first tRNA is free to leave the ribosome. Thus, as the ribosome reads down the mRNA, the chain of amino acids becomes longer and longer until, at the end of the mRNA molecule, the protein is complete. The ribosome, mRNA, and protein disassociate from one another and go their separate cellular ways.

There are several attributes of the genetic code which are worth mentioning here. One is the redundancy of the code; that is, there are more possible triplet codons (4^3, or 64) than there are amino acids (20). The result is that more than one codon can code for the same amino acid. There are also nonsense codons, which code for no amino acid and serve to terminate a protein. It becomes evident that not all mutations necessarily result in an amino acid change, and that a mutation which produces a nonsense codon from an amino acid specifying codon can wreak havoc far beyond the reaches of many missense mutations which result only in exchanging one amino acid for another, for example, change in the codon GUU to GCU. A deletion or addition is also serious, since it throws the whole message out of phase resulting in a *frame shift*; in this case, the entire message is affected as the result of an event involving a single base.

We are now in a position to define the unit of heredity, the *gene*. A gene is simply a functional unit, a sequence of nucleotides in the DNA, which codes for a single protein. This is an oblique way of stating the *one gene–one enzyme* concept; there are exceptions to this generalization, but they are relatively few. However, *one gene–one enzyme* applies only to that class of genes known as the structural genes, those from which the phenotypic characters arise. There are other functional units, also genes, which repress or activate the structural genes in the maintenance of a coadapted *genome* at any point in the organism's ontogeny; these do not specify the formation of enzymes but, nevertheless, have much to do with the phenotypic expression.

The bacteria, among the simplest living forms, have but one DNA molecule; this is found in the single circular chromosome. Such a chromosome is estimated to have the capacity to specify 3300 different proteins. Much larger and more complex systems occur in eucaryotes, especially among the metazoans and metaphytes. It is evident that the full complement of information necessary for the production of very complex organisms is carried in the gametes that combine to form the zygote. That this can be true seems intuitively baffling because of the extremely small size of the information-carrying portions of the gametes. However, it may be that the potential amount of information available far exceeds the amount which is actually used. Studies have shown that the amount of DNA is moderately constant from cell to cell among the members of a species. Given four different nucleotides and their four nitrogenous bases (*ACGT*), some estimations of the possible number of bits of information can be made. The human sperm, for example, has an amount of DNA equal to about 2 to 3×10^9 nucleotides (that is, 2–3 billion nucleotides). If it is assumed, as is reasonable from available data, that a segment of DNA about 1000 nucleotides long is minimal for the synthesis of a cellular enzyme, then the DNA of a sperm could contain at least 1×10^6 bits of information, and possibly two or three times that much. Theoretically at least, each of the four kinds of nucleotides could occur at any position, so that each of the 1000 nucleotide sequences would have 4^{1000} (or 10^{600}) unique sequences. Finally, from a strictly informational point of view, possible different sperm cells based on these determinations are astronomical—ten to the billionth power (10^{10^9})!

Of course, in a given species such as *Homo sapiens*, the potential variability is strictly limited, conforming to a highly selected set of

VARIATION AND NATURAL SELECTION

characters which are compatible with each other and with the environment of the organism. The extent of potential information in a gamete, as calculated above, is immensely greater than that needed for all the forms of life that have ever existed.

Despite our impressive body of knowledge about genetic and biochemical events within the cell, there are still frontiers to conquer. Indications are that not only is the information content of DNA very great, as we have seen, but also the kinds of regulating mechanisms are elaborate and probably involve several levels of control. Some of these points will be brought out in the following sections. More important at present, however, is an understanding of how genes and variability operate within a framework of *inheritance*—the passing of genetic information from one generation to the next.

Some genetics

Long before the field of biochemistry spelled out the letters of the genetic code, geneticists had discovered many of the fundamental properties of inheritance by conducting breeding experiments and observing changes in phenotype from generation to generation. The most important single fact that has emerged from these experiments is that inheritance is particulate, and not blended as was thought in Darwin's time. Inheritance is extremely complex even for single characters, but the fact that characters are not destroyed or blended with other characters in the course of the recombination that accompanies sexual reproduction gives us a method of analysis which has proven very useful in constructing models of genetic mechanisms.

For illustration, we will examine an extremely simple case of inheritance, with the understanding that more elaborate circumstances generally pertain. We will assume that a character is controlled by a single pair of genes, one on each of the homologous pair of chromosomes (one from the male parent, one from the female). Each gene is called an *allele* for the character under consideration. We will further assume that one of the alleles is *dominant,* and that it therefore will be expressed in the phenotype when it constitutes either one or both members of the allelic pair. The other allele is, by definition, *recessive;* that is, its gene product will not be expressed in the organism's phenotype if the dominant allele is present.

Suppose now that we have a species of flower which manifests only two colors, red and white. This is actually the situation with which Mendel dealt in his pioneering genetic experiments in the mid-nineteenth century. Let us say that red is the dominant color and white the recessive. We will assign the symbol A to the dominant gene and a to the recessive one. One or the other of these alleles will be carried in each haploid gamete. These products of meiosis will fuse to form a zygote whose genetic constitution with regard to these particular alleles can be illustrated by a diagram (fig. 26). Note that each dia-

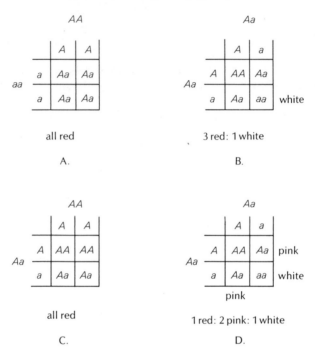

Figure 26 Simple diagrams of the ratios of types of allelic pairs of genes resulting from different crosses. Allele A is dominant and allele a is recessive in diagrams A, B, and C. A produces a red color in the phenotype whenever it is present. a will result in a white phenotype only when both alleles are a, that is, the pair is homozygous, aa. In D, there is incomplete dominance of A, so that in the heterozygous, Aa, a pink color results, and red or white occurs only in the case of homozygotes AA or aa.

gram represents all the possible combinations of genes in the offspring of the particular parents, and that the ratios of genotypes actually represent the probability of each offspring having a given genotype.

Whenever an organism gets two doses of the same allele (AA or aa in our example) it is said to be *homozygous* for that gene. Where a combination of alleles occurs (Aa), the individual, regardless of

VARIATION AND NATURAL SELECTION

phenotype, is *heterozygous*. In the example of the red and white flowers, whenever *A* is present, the resulting flower will be red. White flowers can only result from an *aa* homozygote.

Unfortunately, the great majority of genes are not simple dominant-recessive systems. In many cases, codominance or incomplete dominance occurs (fig. 26D) in which phenotypic (but not genotypic!) blending of characters occurs. In such a system, for example, pink flowers might occur, and several shades of pink at that. Still, from breeding experiments, it is possible to trace alleles through successive generations.

There are even more complexities, but we need only state a few for our purposes. The most common phenomenon is *pleiotropy*, in which one gene exerts its effects on more than one phenotypic character. On the other hand, there are many single characters that are influenced by more than one gene, these characters are *polygenic*. The number of possible allelic combinations involved in a two-gene polygenic system, assuming that both genes are governed by simple dominance, rises from two phenotypic classes with three genotypes (*AA,Aa, aa*) to four phenotypic classes and nine genotypes (*AABB, AABb, AaBb, AaBB, AAbb, Aabb, aaBB, aaBb,* and *aabb*). A more limited phenomenon is the existence of sex-linked characters; these are controlled by genes on either the *X* or *Y* chromosomes. The disease haemophilia is caused by a sex-linked mutation on the *X* chromosome.

Genes, as we noted in the previous section, are capable of interacting with one another as activators, integrators, and suppressors. As with structural genes, one integrator may affect many genes, and one gene may be affected by a number of integrators. The significance of such a multiplicity of complex systems of control and expression is that it results in a coadapted genome; it also places stringent limitations upon what kinds of mutations are tolerable and, thereby, limits the variability within a given species.

The whole matter of gene functions is being intensively studied in analyses of the evolution of genomes (all genes in a cell) of multicellular organisms as they relate to the origin of complex genetic systems. Equally intense efforts are being made to solve the related problems of the effects of genes on the development of individual organisms and how their interactions result in the differentiation of cells and tissues in the course of ontogeny. These are critical and fascinating fields in which rapid progress is being made and from which deeper insights into evolutionary processes are emerging. From all such studies, the genotype proves to be the critical entity both in ontogeny and evolution, with the genes providing the particulate basis of inheritance as well as operational unity deriving from integrative and regulatory processes.

Changes in the genotype

We have stated that mutations at the nucleotide level are the ultimate source of variability in populations. However, there are phenomena which can affect entire chromosomes or major portions of them, and these, too, may be reflected in the organism's phenotype where they may be acted upon by natural selection.

There are many types of chromosome alterations, but they fall into two categories—positional and multiplicational. *Positional* changes involve exchange of materials between homologous chromosomes (*crossing-over*) or nonhomologues (*translocation*), or even between cells (*transformation*) (fig. 27). They also include *fragmentation* of single chromosomes, *fusion* of separate ones, and *inversions* of sections within, or *deletion* of material from individual chromosomes. These changes serve to rearrange the genome and to change the relationships of both structural and integrative genes to one another.

Multiplicational changes are thought to have played as important a part in evolution as have positional changes. In plants, a potent mechanism of evolution involves duplication of the entire diploid set of chromosomes so that the adult plant has twice as many chromosomes as it would normally have. Such a plant is called a *tetraploid*; if several duplications occur, the condition in general is called *polyploidy*. The duplicative event is spontaneous, and can create new species of plants "overnight" whose phenotypes are quite different from those of the diploid parents. Polyploidy is rare in animals, but it may have been involved with increasing the number of chromosomes in the genomes early in the evolution of many groups of metazoans.

Figure 27 (*opposite page*) Four types of chromosomal mutations diagrammed to show how they occur. The different parts of the arms in the pair of chromosomes are indicated by different patterns so that they may be followed through. The circular structure to which the arms are attached is the *centromere*. A, An *inversion* in which the order of chromosomal material is changed. B, A *crossover* in which parts of one of the arms of each of the two chromosomes are exchanged C, *Translocation* in which parts of an arm of one chromosome become attached to an arm of the other of the pair. D, Deletion, in which a part of the arm is lost.

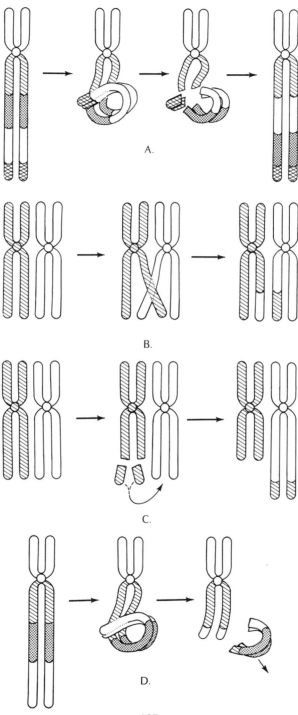

Another phenomenon appears to have equal impact on plants and animals, and that is gene *duplication*. The actual significance of gene duplication is the subject of debate at the present time, but it is thought that extra copies of genes that are more or less permanently *turned off* can accumulate mutations which can, at a later date, contribute to the integrated genome. Whether this is completely true or not, it is evident that there is much duplicated material in chromosomes, and that a good deal of it may not be used.

As important as the above changes may be in the long-term evolution of groups with major differences, production of variability within populations as expressed throughout the gene pool arises primarily from recombinations of existing alleles by sexual reproduction. The distribution of maternal and paternal chromosomes in each gamete is random, and the variability produced by the union of two such gametes from different individuals is expressed in the offspring. Thus, mutation need not be the immediate source of variability; quantitative changes in gene frequency can accrue from selection upon differences generated by recombination alone. Also, there is no necessary proportional relationship between evolutionary rates at the species level and the rates of various kinds of mutations. Mutations in this context are seen as a genetic resource upon which populations can rely if there is selection in their favor, but which are not at the forefront of events in an actively evolving population.

Natural selection

Natural selection is the directional agent in the evolutionary process; it acts upon individual organisms to effect their differential survival and reproductive success. At the population level, selection becomes a statistical phenomenon favoring certain alleles and eliminating others. The result is that the gene pool either stays within limits prescribed by selection or it evolves as the proportions of its component alleles change relative to one another. Only that portion of the gene pool which is phenotypically expressed is under direct selection, and the same characters are not expressed or selected at all stages of ontogeny.

One of the most interesting and difficult areas of study in biology is the analysis of how genetic information comes to its final expression in the adult phenotype. Evolution, in a developmental sense, proceeds by a process of changing ontogenies, so the results of such studies have obvious implications about the nature of natural selection. For many years, the most fruitful avenue of approach was through experimental embryology; the data amassed as a result of these studies is now being augmented and elucidated by studies cast at the molecular level. It is beyond our scope to deal with these problems in detail, but it should be borne in mind that all parts of the population—embryonic, larval, and adult—are subject to natural selection and that different phases may have vastly different selective pressures imposed upon them.

It is useful to begin looking at populations from an *ideal* condition, in which there are many thousands to millions of individuals, no immigration or emigration, and random interbreeding with neither mutation nor selection. In such a case, the population is said to be in *equilibrium*, and its gene pool will not change with time. This is a verbal expression of the mathematical *Hardy-Weinberg law*. As might be expected, these conditions are not fully met by any populations that exist over considerable periods of time. If that is true, and if evolution is defined as change in the genetic makeup of successive populations, then evolution is a normal characteristic of populations. The importance of this concept cannot be overstated.

In small populations, there is a relatively great likelihood that chance changes in gene frequencies will occur; that is, it may be possible for an allele to penetrate such a population without the assistance of natural selection. This is the phenomenon referred to as *genetic drift*. Its importance as a factor in evolution is still a matter of debate, but it is evident by the Hardy-Weinberg equation. There are some populations which are large enough and isolated enough that the ideal condition might be approximated. Nonrandom breeding and mutations do occur, however, and these ultimately will upset equilibria.

Mating patterns have played important roles in evolutionary events at the population level. The three patterns shown in figure 28 illustrate ways in which this occurs. Fully random mating over the entire population, the situation in which each of its members has equal chances of mating with any other of the opposite sex (fig. 28A), rarely, if ever, occurs. If significant departure from randomness is present, the probability that the Hardy-Weinberg ratios will pertain is very low. In this event, natural selection may become differentially effective in different parts of the population. This situation is shown in patterns *B* and *C* in figure 28.

Any external influences upon a population will tend to shift its equilibrium by giving selective advantage to some alleles over others. As a simple example, we may take the use of DDT on mosquitoes. In a population of mosquitoes, there are almost always some phenotypes

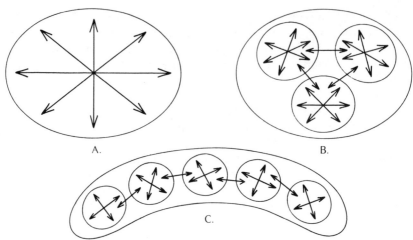

Figure 28 Three breeding patterns shown by diagrammatic drawings. The arrows indicate the range of breeding and the outer circles give the limits of the species population. The inner circles give the limits of subspecies populations. A, completely random breeding throughout the whole population; B, three subpopulations with random breeding within, but limited breeding between; C, a series of subpopulations along the cline of the species, with random breeding within but limited breeding between, and this only between adjacent subpopulations.

resistant to DDT. If all individuals in the population were exposed to the spray, which is unlikely, then the next generation would consist only of the progeny of immune types. Not all of these offspring would necessarily be resistant, but a drastic increase in the percentage of resistant types is assured. Even if the entire population were not exposed to the chemical, the advantage of the resistant forms would be significant enough to result in the accumulation of the favored genes over a period of generations.

Another example of the effects of natural selection concerns the development of industrial melanism in various invertebrates in England, particularly in the peppered moth *Biston betularia*. Studies by the British entomologists E.B. Ford and H.B. Kettlewell correlated the increase in industry in certain areas with the darkening of surfaces used as resting places by the moths, which have a genetically determined light and dark phase. Cryptic coloration of the light phase made the individuals possessing it virtually invisible to predators when they were resting on normal bark (see fig. 36, chapter 7). Initially, as expected, the light phase predominated in the population, but as the trees became darker, the advantage of obscurity shifted toward the darker phase, and the genetic composition of the population altered until, finally, the dark phase became dominant. This was a selective process that acted upon the phenotype and thus altered the genotype in favor of the genetic composition that produced darker moths.

The processes of natural selection have been observed in actual populations, but they also have been subjected to rigorous and extensive mathematical analysis by such students as Sewall Wright, J.B.S. Haldane, and Ronald Fisher. In the course of their work, they established unequivocally the nature and effectiveness of natural selection in evolutionary processes. The very important point that has emerged is that even an extremely slight selective advantage will result eventually in the accumulation of the favored gene in the gene pool of the population.

The sorts of changes we have considered so far are creative and contribute to evolutionary change in a positive way. Natural selection, however, has a conservative side. We have noted previously that a large part of the accumulated genetic variability may be deleterious, and would reduce the fitness of the phenotypes in which it was expressed. Some genes are even lethal; that is, they result in the death of the individual at some stage of development. These deleterious genes can be carried in the genotype as long as they are not expressed phenotypically; however, when they do appear, perhaps as homozygous recessives, they are relentlessly weeded out by selection. This tends to reduce the proportion of the offending genotype in the population. Thus, through its continual mitigating action, natural selection has brought populations to a fine state of balance with the environments in which they exist. Any allele that enhances this balance will be selected for; any that disturbs the balance will be selected against.

The phenotype is a compromise of all the selective forces, internal and external, to which it is subjected in the course of its prereproductive and reproductive lifetime. Although analyses are often made on a character-to-character basis for simplicity, the selective value, or fitness, of a character can be understood fully only in its relationship to all others; to consider it otherwise can be misleading.

In the preceding chapter, we noted the significance of *key* characters in the evolution of mammals from reptiles. Endothermy in mammals, basic to a number of mammalian traits, is a case in point. It was said that the existence of one such key character, selectively advantageous in its own right, provided circumstances which enhanced the selective advantages of certain other characters. This is a case of interdependency seen at a very gross level in the fossil record. However, similar phenomena have been observed in laboratory organisms, particularly in the fruit fly *Drosophila*. The existence of interdependency at microevolutionary levels is critical, because it is at this level that the *processes* seen at grosser levels actually have their origins.

A well-documented example of interdependency of characters is found in the breeding of domestic animals. Emphasis upon such features as egg-laying in fowl or meat production in cattle is accompanied by loss of adaptive effectiveness in many other characters. Few domestic animals can compete with their wild relatives in which the balance of characters has been preserved by the conservative action of natural selection.

Populations respond to environmental changes by utilization of their variability through the agency of natural selection. Survival of successive populations depends upon the variability they carry. If stresses exceed the capacity to adjust, extinction of the population will take place. If the variability is such that it includes features suiting members of the population to different environments, shifts to these new circumstances may be made as appropriate environments become available.

Evolution takes place through the continuing interplay of natural selection and variability. The very minor changes in genotypes and phenotypes apparent in populations from season to season, year to year, and decade to decade lie at one end of the scale of observation. The origin of new kinds of organisms and adaptive radiations lie near the other end. But both of these phenomena, and all the intermediate processes, are part of the larger phenomenon of organic evolution and the origins of diverse organisms on the face of the earth.

Summary

Species have been defined in many ways in the past. Currently, the commonly used definition is based on population concepts. A species is a group of interbreeding populations that are reproductively isolated from all other such groups. Many criteria are used to differentiate species—morphological, behavioral, cytogenetic, hybridizational, and biochemical. In evolution, it may be the species as a whole or only some part of it that is the basic unit of evolutionary change.

SUMMARY

All species populations have great variability, both expressed and unexpressed, in the phenotype. It is upon this variability that natural selection acts in its role of directing evolution. Variability arises as a result of mutations and rearrangements in the information-bearing part of the cell, DNA. The informational potential of DNA of the cells is infinitely greater than that necessary for coding in replication and reproduction of even the most complex organisms.

In populations, variability is maintained and spread by recombination of alleles through the processes of sexual reproduction. Phenotypic expression of the genotype depends upon particulate inheritance of alleles and the dominance relationships of particular alleles. Structural genes directly influence the phenotype by their role in coding of enzymes through the agency of RNA. Various modifier genes act to influence the action of structural genes, acting differentially at various stages of ontogeny of the phenotype.

Natural selection alters the proportions of genotypes in successive populations. If no selective factors exist in large populations, a persistent balance between genotypes will be attained. This is rarely completely attained in nature, although populations often approximate the balanced ratios of alleles predicted by the Hardy-Weinberg law. This being so, change will usually occur, and evolution is the normal course of events for populations through time.

Rates of change depend on many factors, in particular upon the breeding structure and size of populations and the stability of the environment. Generally, natural selection is conservative, tending to eliminate innovations. With time and changing factors, slow, conservative change can give rise to basic changes of the sorts seen in a broad overview of the course of evolution.

IMPORTANT CONCEPTS

General Terms

ALLOPATRIC: describes species or populations that form parts of separate communities and so are not able to interbreed.

ISOLATION, REPRODUCTIVE: the condition prevailing when closely related organisms are unable to interbreed because of spatial or temporal separation or genetic or morphological incompatibility.

NATURAL SELECTION: the differential action of the environment upon the survival and reproductive success of individuals within a population or species. Gives evolution its *direction*.

POPULATION: any group of entities identified by some common aspects.
- LOCAL: a community of potentially interbreeding individuals living in a given, restricted area.
- SPECIES: all the individuals in a given species.
- INTERSPECIES: several interacting species.

SPECIES: a distinct sort or kind of something.
- GENERAL BIOLOGICAL: a distinct kind of organism having certain distinguishing phenotypic characteristics.
- DYNAMIC: a group of interbreeding natural populations that are reproductively isolated from other such groups (Mayr 1970).
- TEMPORAL: a lineage (an ancestral-descendant sequence of populations) evolving separately from others and with its own evolutionary role and tendencies.

SYMPATRIC: describes species or populations that are part of the same community whether or not they are capable of interbreeding.

TYPOLOGY: in biology, a method of taxonomic study in which a single individual or group symbolizes the properties of a more inclusive group. For example, each species has a type individual, each genus has a type species, etc.

VARIATION: the totality of genetically controlled differences in the phenotypes of all individuals with a population or species.

Genetic Terms

ALLELE: the particular expression of a gene, or any of its alternative expressions throughout the population of organisms under consideration.

CHARACTER: any measurable or definable component of an organism's phenotype.

IMPORTANT CONCEPTS

CHROMOSOME: a DNA-containing body found in the nucleus of eucaryotes and bound to the cell membrane of procaryotes that is responsible for cell replication. It carries the genes.

CLINE: a character gradient or *slope*. Usually applied to a change over a geographical range. Also applied to a temporal range (*chronocline*).

CODON: a nucleotide triplet of bases that specifies a particular amino acid.

CROSSING-OVER: an exchange during meiosis of corresponding segments of homologous chromosomes.

GENE: a unit of inheritance; a sequence of base pairs on DNA specifying particular structures, functions, or enzymes.

GENE DUPLICATION: the formation, by whatever mechanism, of multiple copies of the same gene; not all of these need be functional at a given time.

GENE FLOW: the spread of genes through natural populations by means of the dispersal of gametes.

GENE POOL: the totality of genes in a given population at a given time.

GENETIC: pertaining to origin or genesis.

GENETICS: the study of heredity and inheritance in organisms.

GENETIC DRIFT: random changes in gene frequencies in the population from one generation to the next. More pronounced in small populations.

GENOME: pertains to the hereditary materials in a single individual only.

HARDY-WEINBERG LAW: a mathematical generalization describing gene frequencies in an infinitely large population whose members interbreed randomly and upon which no differential selection occurs. The prediction is that the gene frequency will not change with time. Can also be used to predict gene frequencies away from equilibrium.

HETEROZYGOUS: having different alleles at the same loci on two homologous chromosomes.

HOMOZYGOUS: having the same alleles at the same loci on two homologous chromosomes.

HYBRIDIZATION (OF DNA): procedure in which separated strands of DNA from one organism are mixed with DNA from an organism from a different taxon. The extent of the combining of the two is a measure of the similarity of base pair sequences, and thus is a measure of the genetic similarity of the groups.

MELANIC: dark, black.

MUTATION: a change in the genetic material. May be used to refer to a change in phenotype or to any sort of change in the genes themselves, including crossing-over, translocation, and other whole-chromosome phenomena. The common use and the one in this chapter refer to change of a single gene involving additions, deletions, or alterations in one or more base pairs of a gene.

PHENOTYPE: the totality of characteristics of an individual. Includes all stages of ontogeny and adulthood.

PLEIOTROPIC: a gene that affects several characters of the phenotype.

POLYGENIC: a character that results from the action of several genes.

POLYPLOIDY: the duplication of the entire diploid genome (2N) any number of times; can occur spontaneously, and may become the stable condition in some species of plants.

TRANSCRIPTION: the process of transferring genetic information from DNA to RNA.

TRANSLATION: the process of converting sequences of codons on RNA into sequences of amino acids in a protein.

TRANSLOCATION: the shift of part of one chromosome to another chromosome.

REFERENCES

Clausen, J.; Keck, D.; and Hiesey, W. 1948. *Experimental studies on the nature of species III. Environmental responses of climatic races of Achillea.* Carnegie Institute of Washington Publication 581.

Dobzhansky, T. 1951. *Genetics and the origin of species.* 3d ed. New York: Columbia University Press.

Handler, Philip, ed. 1970. *Biology and the future of man.* New York, London, Toronto: Oxford University Press.

Mayr, Ernst. 1970. *Populations, species and evolution.* Cambridge: Harvard University Press, Belknap Press.

Simpson, George G. 1961. *Principles of animal taxonomy.* New York: Columbia University Press.

Watson, J.D. 1970. *Molecular biology of the gene.* 2d ed. New York: W.A. Benjamin, Inc.

7

Examples of evolution in action

Perspectives

This chapter treats some specific examples of evolution drawn largely from the current scene. These examples are designed to illustrate at the phenotypic and population level the kinds of changes for which mechanisms were discussed in the preceding chapter. The items in the first part are somewhat general, involving three more or less distinct patterns of evolution which can be seen both in living populations and in the fossil record. The remainder are more specific and illustrated by cases drawn from recent work on modern populations.

It is a tenet of Neodarwinian theory that changes such as those portrayed by these various events form the fabric of evolution. Such small changes, by continuing accumulation of the resultant differences in successive populations, are considered to have led to broader more pervasive evolutionary events which have been described in the first chapters of this book. It is not demonstrable that this suggestion is in fact the case, nor can it be shown that it is not. There is strong evidence for an inference that such small changes are at the heart of all or most all evolution, and this assumption is made in this chapter. A bit of skepticism, however, is always useful.

Most of the events used for illustration portray patterns and processes that are readily explainable by evolutionary theory. Only a few show evolution in progress. Natural events in which the actual course of change can be observed are unusual, first because of the slow rate of change and second because of the extremely difficult task of measuring such small changes. This measurement requires extensive sampling of populations, which not only may be very difficult, but often results in disruption of the structure of the populations being studied. More and more experiments of this type are being carried out, and progress is being made. The results often seem paltry compared to the expenditure of time and energy devoted to gathering that data.

Most of the observations in which evolution is seen in action have been made either in the laboratory or in the field, where man has consciously altered circumstances so that modifications have occurred more rapidly than would otherwise have been the case. It is, of course, clear that in breeding of domestic animals and plants, an increase in the rates of evolutionary processes has been brought about. This sort of acceleration was important to Darwin in the formulation of his ideas of natural selection, but both he and Wallace leaned very heavily upon what they observed in natural populations and upon the consistent success of the concept of natural selection in formulations of explanations.

Divergent, parallel, and convergent evolution

The three patterns of change described by these terms are illustrated in figure 29. The diagrams and explanations will clarify as well what is usually meant by the terms homology and analogy, which enter into all discussions of divergence, parallelism, and convergence.

Divergence is the normal evolutionary course as races, subspecies, and species form from parent populations. Isolation of parts of the interbreeding parent populations by one means or another is necessary if divergence is to occur. As a rule, divergence is accomplished by geographic isolation which prohibits interbreeding of individuals of two or more portions of an originally continuous population. Mountain ridges, rivers, islands, and similar features may contribute to such isolation. Figure 30 shows a classic example of divergence based on Darwin's finches on the Galapagos Islands. As they reached these islands, the different elements of the parent population diverged into different habitats and developed different habits. A large array of species was the result.

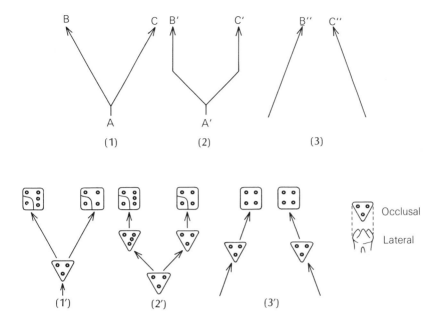

Figure 29 Divergence, parallelism, and convergence, diagrammed in upper series and shown as applied to mammalian teeth in the lower series. Also, in the lower diagrams, the difference between analogy and homology is illustrated. (1) Species A gives rise to two *divergent* lines which lead to species B and C. (2) Species A' gives rise to two lines which diverge and then develop in parallel. B' and C' are similar in many respects, but not in all. (3) Species B" and C" evolve to resemble each other more closely in phenotypic characters. Their evolution is *convergent*. (1'–3') diagram of teeth of mammals in occlusal view (see inset) following the courses shown in the upper diagrams. (1') shows a divergence. Two *squared-up* teeth come from a common ancestor of triangular shape. Because the new cusp was not present in the common ancestor, it is considered *analogous* in the two derived teeth. The triangle of the ancestral tooth has three cusps. Each is present in the derived teeth. Each cusp is homologous with its counterpart in the derived teeth, because it is present in the common ancestor. (2') A diagram of teeth comparable to the diagram in (2) above. Homologies and analogy as in (1'). (3') Species B" and C" are from different ancestors. None of the cusps was present in any common ancestor. Under this assumption, none of the cusps is *homologous*, i.e., all are *analogous*.

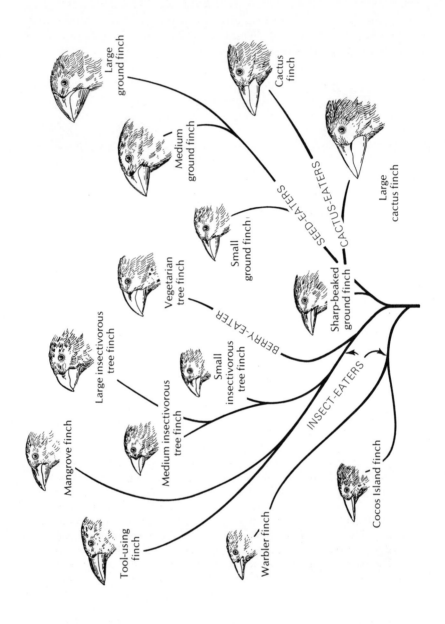

DIVERGENT, PARALLEL, AND CONVERGENT EVOLUTION

In phyletic speciation, where in time one species leads directly to a subsequent one, divergence does not enter in. Time, rather than space, provides the genetic isolation. How common this process is remains an open question.

Parallelism occurs when two or more populations, after divergence from a common stock, evolve in much the same way, developing similar properties. This results from like responses to similar environments after only slight to moderate divergence from the parent population. The two or more separate lines generally have retained very similar genotypes. When lines from very different genetic backgrounds become increasingly similar by adaptations to similar circumstances, the process is termed *convergence*.

Divergence is a term with a clear meaning, but parallelism and convergence are used rather loosely and with considerable overlap. The wings of insects and birds (fig. 31), for example, are clearly convergent, because they represent somewhat similar structures used for the same purposes, but coming from two backgrounds in which there was no common ancestor possessing wings. These wings are *not homologous*. Wings of bats and birds, likewise, are usually considered convergent, because there was no wing in their common ancestor. Again, they are not homologous. But here the structural base, the forelimb, was present in the common ancestor. The limbs *in toto* are *homologous*, whereas the wings derived by their modifications are *analogous*. In a very broad sense, bird and bat wings might be considered to represent parallelism rather than convergence. Common usage of the word parallelism does not, as a rule, include the extent of divergence exhibited in this case, but this usage is not always followed.

As for many biological terms, homologous and analogous, parallel and convergent, are useful but poorly defined. Confusion could be eliminated by careful definition, but this has not been done with enough success that any definitions have gained full acceptance. Irrespective of terminology, questions of whether structures in different categories of organisms do or do not have a common morphological or genetic base (i.e., are homologous or analogous) have a critical bearing upon interpretations of their evolutionary history. To clarify whether or not a common base exists, we will look briefly at a few instances at rather gross levels. At lesser levels, the subtleties may be greater and solutions more difficult.

Figure 30 (*opposite page*) Adaptive radiation at the species level involving Darwin's finches from the Galapagos Islands. Note the diets and the beak forms adapted to different feeding habits. Data from D. Lack (1947).

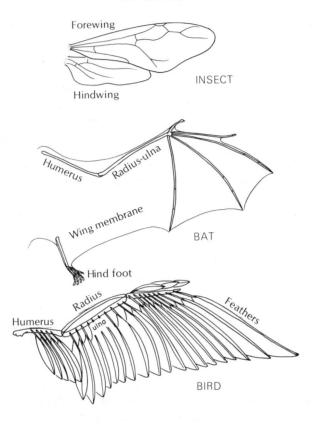

Figure 31 Homology and analogy. The three wings—insect, bat, and bird—are structures which have similar use and generally similar form. Wings, however, were not present in the common ancestors of these three animals, and the flying mechanisms have been evolved independently. Thus, the wings are *analogous,* but not homologous. The bird wing and the bat wing, however, are forelimbs, both derived from reptilian ancestors in which the forelimb was present, with bony elements such as the humerus, radius, and ulna. As forelimbs, then, the two structures in the bird and bat are *homologous.* Furthermore, the humerus in the bat and bird are homologous. Only the wings, as wings, are merely analogous. These two wings and the insect wing are only remotely related, and the wings have formed from completely different ancestral structures. They are in every respect analogous.

Hystrix and Erithizon

Today, animals called porcupines occur both in the New and Old World. They are placed in different genera, *Hystrix* and *Erithizon*, respectively (fig. 32), but have many similarities beyond their super-

Figure 32 Two porcupines—*A*, *Hystrix*, the Old World porcupine; and *B*, *Erithizon*, the New World porcupine. As discussed in the text, these have only a rather remote common ancestor, but they have come to be very similar. Either they may be considered to represent an extreme condition of parallelism, or, more usually, a case of convergence.

ficial possession of quills. The masseteric musculature of the jaws, in particular, show strong similarities. The patterns of jaw musculature, especially the masseteric, have long been used to distinguish the major groups of rodents: primitive rodents, mostly extinct; squirrel-like rodents; a group including, among other kinds, rats and mice; and the hystricomorphs, named from the Old World porcupine. Until rather recently, the last group, the hystricomorphs, was considered to include both Old World forms and a host of New World forms, mostly South American, but including the North American porcupine.

It is well known, however, that the continent of South America was isolated from North America, as well as from the rest of the world for a considerable part of the Cenozoic era. Now it seems clear that the

North Amerian or, better, the New World porcupine arose from South American ancestors. The morphological patterns distinctive of the hystricomorphs as a whole thus appear to have come into existence well after the separation of North and South America had occurred. No migration on land from the Old World to South America by way of North America was possible. *Erithizon* came to North America from the south only when land connections had reformed. Either, then, the common patterns of jaw musculature arose independently in the Old and New Worlds (by convergence or parallelism) or there was some sort of connection between the two areas after the hystricomorph pattern of the musculature had evolved.

What has happened in efforts to resolve this problem offers a good case history of how opinions may change in fairly rapid succession. For many years students of the rodents went to great lengths to explain how the common pattern of the masseteric muscles could be homologous in the two groups. Explanations included suggestions of land bridges across the South Atlantic Ocean, migrations through Antarctica, and rafting across the ocean. During this period, it was firmly believed that the continents had maintained their present positions throughout the history of the world. More detailed study of the rodents and the seeming impossibility of migration between Africa and South America led to the conclusion that the rodents of the two continents must have developed independently from a primitive rodent stock. The two groups once united as hystricomorphs were divided into separate suborders in most classifications. The two porcupines thus had long independent histories and their resemblances were the result of massive convergence.

Many students still hold to this solution. Recently, however, the concept of continental drift has been supported by a great deal of evidence. The relative positions of continents appear to have changed, and Africa and South America may have been much closer together about 25 MY ago. The idea that rodents may have passed between the continents seems much more feasible, and the hystricomorphs may actually be a single group with a common hystricomorph ancestor. Still, of course, the Old World and New World porcupines have had a moderately separate history during many millions of years, but they developed their similarities from a common base and these similarities can be explained merely as parallel evolution, not as convergence. Today, this idea is gaining much support.

North and South America

The relationships between these two continents, as we have seen, are classic among reconstructions of evolutionary histories and interrelationships of the biotas of the two continents. For comparisons, a simple index of similarity may be used: $s = (C/N_1) \times 100$, where s is the index, C the number of taxa common to each area, and N_1 the number of taxa in the larger sample. Today this index between North and South America is $19/41 \times 100 = 46$, based on families of mammals. The table on page 167 shows the families involved. Had this index been taken 15 MY ago, it would have been much lower.

North and South America became separated about 60 to 65 MY ago. Except for a few highly selective interchanges, isolation continued until about 10 MY ago. During this time, a mammalian complex of placental herbivores and marsupial carnivores developed in South America, in contrast to the fully placental mammals system in North America. Many of the carnivores and herbivores of South America and their North American counterparts show strong convergence. Horselike, camellike, rodentlike, and even elephantlike herbivores were present (fig. 33) in South America. Among the marsupials was a remarkable sabre-toothed "tiger," which was very like its placental North American counterpart, figure 34. In addition, however, very different animals developed. Among them were the New World monkeys, paralleling those of the Old World but coming from a very different base among the primates. Unlike animals anywhere else, except for a few convergent Old World genera, were the edentates. The sloths, armadillos, anteaters, and glyptodonts are examples.

When North and South America were joined about 10 MY ago, faunal mixing began, with northern animals pushing southward and southern forms pushing northward. The parallel and convergent types became competitors. For the most part, the northern carnivores and herbivores eliminated their South American counterparts. The rodents and monkeys were little affected. Rodents made some progress north and are represented by the porcupines in North America. Edentates, with little direct competition in North America, moved north en masse and successfully established themselves as part of the North American fauna. Many died out in the late Pleistocene, only a few thousand years ago, just as they did in South America. Both continents suffered a major extinction of large mammals at that time,

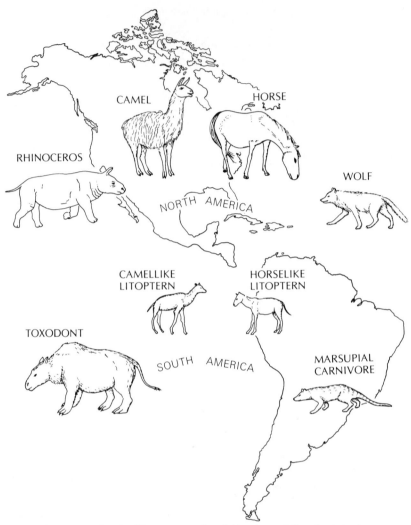

Figure 33 Look-alike mammals of North and South America, developed independently from remote ancestors on the two separated continents. Note that the South American carnivore is a marsupial, and the North American (wolf) is a placental. All herbivores, however, are placentals. See the text for discussion.

and the coincidence with the invasion of the New World by man has been taken by some to indicate that man may have been an important factor in this extinction. Horses which had successfully invaded

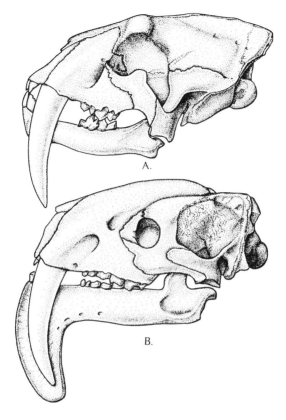

Figure 34 A sabre-toothed tiger (*Smilodon*) from North America and a marsupial sabre-toothed carnivore (*Thylacosmilus*) from South America. These two represent a strong evolutionary convergence to a similar form and probably a very similar way of life.

South America from the north died out, just as they did in North America, but camels have remained in South America (llamas).

Australia and Asia

Another, and also classical, case that we will examine briefly involves the relationships between Asia and Australia. Australia was isolated from the rest of the world some 65 to 70 MY ago. By some means, not fully understood, only marsupials populated the area at the time of isolation. They evolved in such a way that many of the roles played by placental mammals on other continents were carried out by marsupials in Australia. As in South America, there were marsupial carnivores, some catlike and some doglike. Also there were rodentlike forms, even flying phalangers which resembled flying squirrels. Many insect eaters were present, one of which resembled the placental mole closely. Other animals, such as the kangaroo which has roles not unlike those of deer and antelope, looked little like their placental counterparts. Both morphological and environmental convergences are striking.

Some placental mammals—bats, rats, and dogs—reached Australia prior to the time that European man arrived. It is primarily these animals that give the Australian fauna an Eurasian complexion. If bats are included, the similarity index based on mammalian families for Australia and Asia (including southeast Asia) is $8/36 \times 100 = 22$. If bats, which as fliers can easily move between Australia and Asia, are eliminated, the index is about 5, and this includes murids, rats, and the dingo dog. The last animal probably was introduced by man. If Australia is compared, as it stands today, with South America, the similarity index is 7, or without bats, about 2. The long isolation of the continent shows up graphically in these figures. It is under such circumstances of isolation that remarkable convergences affecting many parts of a fauna develop.

Mimicry

The familiar brown and black Monarch butterfly, common throughout the United States, is distasteful to the birds which prey on butterflies. The Viceroy butterfly lives over part of the range of the Monarch and is palatable to birds. It resembles the Monarch closely in form, color, and markings (see plate V). The flight patterns are somewhat different, but the two are sufficiently close that reduced predation upon the Viceroy results from its resemblance to the noxious Monarch.

The poisonous coral snake with its highly colored ring-banding is matched in general appearance by several nonpoisonous snakes over its geographic range. The selective advantage of the poisonous snakes appears to carry over to the nonpoisonous look-alikes.

These and other similar cases throughout the animal kingdom represent what is called *Batesian mimicry*. The Monarch and the coral snake are *models*. The model has some feature that gives it selective advantages in its ecological situation. The *mimic*, such as the Viceroy, has evolved to look like the model by virtue of the selective advantage which this confers on it.

Interpretations of mimicry are full of hazards. First, of course, the evolutionary events which have produced the resemblances have not been seen but are merely inferred. Experiments have demonstrated that mimics do have selective advantages, but the evolutionary events that have presumably led to the resemblances have not, of course, been duplicated. The time available in experiments is simply too short. It is important, too, to remember that the models and mimics are seen through the eyes of man who can only make inferences as to how the organisms involved react. Just because a Monarch tastes bad to man does not mean it also tastes bad to birds; but the actions of birds suggest that this is the case. We cannot be sure that banded coral snakes and others that look similar to us also look similar to natural predators. Such difficulties are not insurmountable, but all cases must be scrutinized with care. Some of the many supposed instances of mimicry have been carefully studied and substantiated, but many others have not.

Mimicry may extend far beyond a pair of species to include several in a mimetic complex. These complexes may include members widely separated taxonomically. A good example is found in the *Lycus* mimetic complex, figure 35. The model is the beetle *Lycus fernadzi*, which appears to be repugnant to predators. Mimics are found among other beetles, wasps, and sow bugs. Primarily, the *Lycus* complex exhibits Batesian mimicry. Some of the mimic beetles, however, are themselves noxious.

The more limited *Lycus* complex, which includes the suites of noxious forms, represents another type of mimicry known as *Müllerian*. In this type, the model and mimics alike have some characters such as bad taste, odor, or appearance which gives them a selective advantage with reference to predators. It has been demonstrated experimentally that predators learn to avoid noxious prey after a small number of feeding trials; thereafter, they tend to reject this prey. If each of the noxious species appeared different to the predator, then each would be subjected to a separate series of learning trials. In a mimetic complex where all look alike, the whole complex serves as the population upon which the trials are made. The number of trials per species is reduced and all members of the complex have the same selective advantage.

Mimicry is a special case of convergence, for the evolution leads to morphological resemblances. It requires, however, that the model

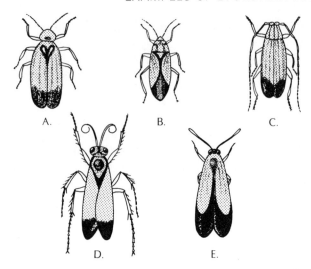

Figure 35 The *Lycus* mimicry complex as discussed in the text. This is a *Müllerian* complex in which the various forms which have come to look like the beetle *Lycus* all are repellent to predators. Common advantage results from the similarities. *A*, the *Lycus* beetle, or soft-shelled bettle; *B*, a bug; *C*, the longicorn beetle; *D*, the spider wasp; *E*, a butterfly. Redrawn from figures in W. Wickler (1968).

and the mimic live in the same region and that the mimic take advantage of some particular characteristic of the model.

Protective form and coloration

Some of the most impelling evidence on the existence of natural selection has come from studies of form and coloring in living organisms. The cases of leaf butterflies and stick animals indicate the remarkable adjustments that can be developed by the selective advantage to be found in concealment. Such concealment is certainly effective with respect to man and, presumably, to predators as well. Very broadly, this is a form of mimicry, but a general one in which the object mimicked is not potentially an acceptable prey. While such cases do not *prove* natural selection, that process is by far the most plausible explanation.

Studies of industrial melanism were discussed briefly in chapter 6. These studies go much farther than mere inference. E.B. Ford developed a model or theory to account for the changes of insects in an

PROTECTIVE FORM AND COLORATION

industrial region of England from predominantly light to predominantly dark. H.B. Kettlewell devised and carried out careful experiments that essentially validated the theory. Populations of various insects in the area have shown this feature of becoming dominated by a dark polymorph. Some, interestingly enough, have not shown this effect, lacking the genetic basis, or light and dark morphs, necessary for the melanization to occur. The moth *Biston betularia* was used in the experiments. This moth has a dark and a light phase, that is, it is *dimorphic*. The phases are genetically determined and subject to Mendelian inheritance.

Original studies dating back to about 1845, before the development of massive industry in England, showed a high preponderance of light individuals in samples of the population. The light dimorph appears to have had a selective advantage over the dark. Over a period of about 50 years, from 1850 to 1900, a shift to about 95% dark forms took place. This accompanied the development of industry which, among other things, resulted in deposits of coal soot on the trees, darkening the resting places of the moths (fig. 36).

The principal predators of these moths are birds, and it was theorized that the light colored moths became increasingly more visible to the birds as the trees became darker. The darker moths, originally at a selective disadvantage, now were better concealed and relatively less subject to predation. They thus contributed more members to succeeding populations and came to be dominant. But, of course, problems arose and questions were posed. It was debated whether the birds, in fact, did prey on resting moths or whether they caught them in flight where shading would make less difference. Dark forms might be distasteful, for example, so that it was a feature associated with the dark form, not the shade itself, that was critical.

Kettlewell's experiments were designed to test such questions. He released large samples of marked moths, in various proportions of light and dark, under differing environmental circumstances. Recovery of moths was carried out by various trapping techniques. Also, by use of field observations, including motion pictures, it was determined that birds did, in fact, prey upon resting moths and that they appeared to do this by sight.

Selection by predators based on shade, light and dark, was demonstrated by these experiments. Important also is the evidence of the rapidity with which a process can occur under strong selective pressures. Of course, there have been some doubts about the validity of the conclusions of the experiments. There always are. And also it will be said that this is not *natural* selection because man had a hand in it.

Figure 36 Industrial Melanism. This is a photograph of the exhibit on industrial melanism in the British Museum (Natural History), London. To the right-hand side are conditions as they are at present, soot-covered. To the left are conditions as they were over 100 years ago. Note the differences in percentages of light and dark moths in cases to the right and left at the top of the picture. Note the differences in visibility of the light moths on the dark and light tree trunks. Published with permission of the British Museum (Natural History), London.

A *natural* phenomenon could produce the same effects. The efficacy of the selective process seems very clearly shown by what has been done.

Polymorphism

In the example of moths just discussed, it was noted that the moth in question had two forms, that is, it was dimorphic. *Polymorphism,* or multiple form, of which dimorphism is a special case, occurs throughout all major groups of animals and plants. It is a term used to describe the situation where discontinuous phenotypes occur in interbreeding populations. Color and shading provide excellent examples because they are readily identified. Blood groups in humans and other vertebrates and right- and left-handed coiling in snails are other well documented cases. Common and usually easily recognized are various types of sexual dimorphism, where males and females have differences not directly part of the reproductive systems. Horns in deer, voices in humans, and coloration and behavior in birds are familiar examples. Less evident, but equally significant, is polymorphism in body chemistry, in enzymes, for example, or in blood groups, as noted above. The effects of polymorphism are important in a variety of kinds of evolutionary events, depending basically upon the different selective values of the morphs.

Polymorphism occurs when several alleles which have distinct phenotypic expressions are present in a population. As a rule, but not always, polymorphic characters are controlled by a few genes. In a very simple case with the alleles A and a, polymorphism results if the phenotypes of AA, Aa, and aa are different. For this to persist in a population, both A and a must retain a high frequency. This condition will be assured if the heterozygote Aa is selectively superior to the homozygotes AA and aa, resulting in selection for both alleles at a higher frequency than would be the case were one of the homozygotes selectively superior.

Polymorphism can be advantageous to populations in many ways. In the case of industrial melanism, it was possible for the populations to adjust rapidly to new circumstances by selection of one of the two types. Different polymorphs of a species may have particular advantages in different parts of the range of populations which cover fairly diverse environments.

Relatively few cases of polymorphism and selection in natural populations have been thoroughly analyzed. With the advent of a technique called *electrophoresis,* which makes the sorting and detection of alleles relatively easy, a great deal of new work is being done. Even so, the amount of work and care for full analysis is immense. A case that has been thoroughly studied is an analysis of several species of the genus *Colias,* the common orange and yellow butterflies, carried out by Ward B. Watt. Throughout the range studied, some 15 species of the genus show polymorphism with respect to pigmentation on the wings, plate V. The species occur over zones, both geographical and altitudinal, in which temperature ranges are considerable. Temperatures are quite high in the south and in the lowlands to the north. They are lower in the southern uplands and in the middle and uplands of the north. The darker representatives, the darker polymorphs, live in the cooler regions. It was demonstrated that the uptake of heat is closely related to the extent of pigmentation of the wings, the darker forms using the sun's energy for heating much more effectively than the lighter forms.

This difference confers a selective advantage to dark forms in the cooler regions where warming is important, but is equally deleterious in warmer regions where overheating tends to develop in the heavily pigmented individuals. The distribution of polymorphs corresponds closely to the pattern predictable from this interpretation.

A much more complex case has been studied by J.R.G. Turner. It involves polymorphism, hybridization, and mimicry in the butterfly *Heliconius* from the Neotropics, especially the forms found in the Guianas (fig. 37). One species, *Heliconius melpomene*, is strongly

Figure 37 A few of the many patterns found in the tropical butterfly *Heliconius*. See the text for explanation. There is a great deal of subtle variation in wing patterns, which are mostly white and red on black. These variations suggest very extensive mimicry and hybridization. Based on J.R.G. Turner (1971).

polymorphic, having some 50 named varieties. Within the species are three monomorphic subspecies. In addition, this species is mimetic in its polymorphism, some varieties copying other monomorphic species within the genus. The members of the genus are all distasteful to birds, indicating this to be an instance of Müllerian mimicry.

It has been argued by Turner that the polymorphism in *H. melpomene* is the result of hybridization of the three monomorphic subspecies. Artificial crossings of these subspecies under laboratory conditions have produced individuals comparable to the natural polymorphs. Much of the variety of pattern is explained by segregation of alleles at four loci, although some unidentified factors do enter in.

This hybridization has been continuing for at least 200 years, as shown by illustrations of butterflies made that long ago. This case, as most of those studied in nature, represents a limited evolutionary event. But it is of such stuff that the totality of evolution is made under the Neodarwinian concept.

Coevolution

This term may be used very broadly to cover interdependency of many species during their evolution, even to the broad level of community evolution. Coevolution is frequently employed in a much more restricted sense to describe reciprocal evolution of a pair of species. Symbiotic relationships fall within this category. The relationships may develop as beneficial to both components or to only one with the other either not affected or affected adversely. Also, evolution within species, such as the development of a social system with morphological differences between castes as with termites, has sometimes been called *social coevolution*.

Most of the instances which have been studied with some care and intensity have been limited and very specific, involving only part of a total system in which they exist. Modelling, based in part on extrapolation from specific cases, has been used to gain an understanding of more complex systems, and analyses of cases suggested by such models have been carried out. These are extremely refractory areas of study, with proper data difficult to obtain. We will look at only a few examples from the many that have been studied.

Plant pollination

Insects, hummingbirds, birds, and bats are pollinators of flowering plants, the *angiosperms*. These plants first appeared during the Mesozoic era and rapidly attained dominance on the continents of the earth. Insects had come into being much earlier, during the middle of the Paleozoic era, but there is no question that they underwent an extensive new radiation as the angiosperms developed. Undoubtedly, insects also contributed to the adaptive radiation of the flowering plants. This reciprocity represents coevolution on a grand scale.

Pollination by mammals and birds arose later, probably in the early Cenozoic; both mammals and birds have had reciprocal relationships with plants in the course of their evolution. The fossil record does not document these broad changes and relationships except in the crudest of ways. To see how they have taken place, we may turn to some analyses which have been made on existing systems.

Flower and insect types Leppik showed a relationship of the levels of insect development to flower development and interpreted it as representing coevolution at a major level over a long period of time. His interpretation is shown diagrammatically in figure 38. That there is a close correlation between insect types and flower types upon which they feed and act as pollinators is clear. In this case, and many similar ones, the mechanism by which selection operates is difficult to determine. Color, scent, and flower configuration are all known to be recognized by insects. Many insects respond to ultraviolet radiation not visible to our eyes. One or more of these conditions may be important, but, except by very careful experimentation for each specific case, determination of which one is not possible. Any such results on living insects and flowers must, of course, be projected back through times past when no possible means for direct analysis existed.

Leppik has proposed that definitive stages in sensory development of the pollinating insects, six in all, are reflected in corresponding levels of the evolution of flowers. The existence of the sensory levels that he has proposed is conjectural and serves as much as a suggestion for direction of research as for an answer to the particular nature of coevolution which existing insects and flowers seem to demonstrate.

Host specificity of wasps W. B. Ramirez has made analyses which show a high degree of host specificity of wasps that pollinate New

Figure 38 (*opposite page*) The relationships between flower types and levels of insect development, as discussed in the text. The ranges of the different kinds of insects are shown to the right as they relate to the kinds of flowers shown and described in the center columns. Redrawn from E.E. Leppik (1957).

ABILITY OF INSECTS TO DISTINGUISH:			Beetles	Flies	Butterflies	Bees	Bumblebees
LEVELS	FLOWER TYPES	CHARACTERS OF FLOWERS					
VI bilateral		Zygomorphic types. Combined characters of IV level, odors and bilateral symmetry.					■
V 3-dimension		Protected types. All characters of III level and third dimension.				■	■
III–IV radiate		Radiate types. Definite numbers, all colors, definite size.			■	■	■
II simple		Simple types. No definite numbers, simple colors, without definitive symmetry.		■	■	■	■
I primitive		Primitive types. Flowers without definite colors and symmetry.	■	■			

World figs. This sort of specificity is common in many types of symbiotic associations, for example, between host and parasite in animals. A specific symbiotic association is of great importance in the control of transmission of various types of disease of both plants and animals. In the present instance, the relationship is vital in the husbandry of fig production. In the case of the figs, each of the approximately 40 species used in the study is symbiotic with one species of wasp which serves to pollinate it. There is one exception in which two species of wasps act as pollinators of a single species of fig.

The wasps can develop only in the gall of the fig flower. Without question, the dual needs of ontogeny of the wasp and pollination of the fig have resulted in evolution involving isolation of both the figs and wasps by the close species relationship which arose as each diverged from its parent species. The fig, *Ficus*, has an immense number of species, about 900, many more than in related genera in which wind pollination occurs. It appears likely that the high number of species is related to the fact of wasp pollination and the host specificity of the pollinating wasps.

Hummingbirds and plant distributions Hummingbirds are pollinators of many species of plants in California. V. and K. Grant have analyzed the relationships of plant dispersal and migratory habits of the hummingbirds, revealing interesting relationships. During their spring northward migrations and their breeding season, hummingbirds are widely dispersed (fig. 39). After breeding and southward migration, the birds tend to congregate in restricted areas in the mountains. The plant species which they pollinate show distributions related to these patterns. The species of spring flowers which the birds pollinate during their dispersed phase tend to occur separately, that is, they are allopatric. The summer-blooming species in the mountains, pollinated when the hummingbirds are aggregated there, occur together in dense assemblages of several species. There seems to be little question that these two patterns are closely related to the migratory and dispersal patterns of the hummingbirds.

Hummingbirds, for the most part, pollinate only plants to which their feeding structures and processes are well adapted. They do, however, sometimes visit other plants and act as pollinators. In areas of dense hummingbird populations where there is strong competition for food, plants not usually pollinated by the birds may form a source of food. Selective factors would tend to favor the plants of a species most subject to visits by the birds and would tend to promote divergence, forming races and new species. This, it is believed, might

COEVOLUTION 157

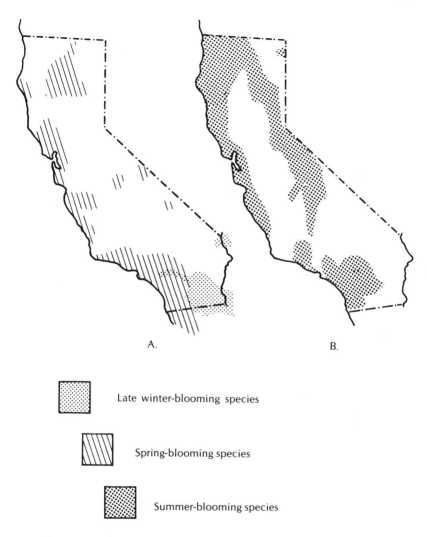

Figure 39 The distribution by seasons of flowering of plant species pollinated by hummingbirds, as discussed in the text. Note the wide spread of summer-blooming species, the spotty distribution of spring bloomers, and the local southern distribution of winter-blooming species. After K. and V. Grant (1967).

lead to the sorts of species aggregates of hummingbird-pollinated plants found in the mountain areas. These include both plants normally pollinated by the birds and others less subject to this in areas where plants are more dispersed. The interpretation, of course, is theoretical—a reasonable deduction, but one which is extremely difficult to test because of its scope and the number of variables.

Larval feeding of insects

P. Ehrlich and P. Raven have made an extensive analysis of the patterns of plant utilization by larvae of butterflies. Many plants have secondary substances which have little to do with their basic life processes. Such substances may be repellant to larvae of most, but not necessarily all, insects. Because of this situation in which relationships tend to be highly specific, important reciprocal relationships have developed between the plants and insects. These relationships have had important evolutionary effects.

A particular species of plant, for example, may undergo a mutation leading to the development of a substance which does not inhibit its normal growth, but which may destroy its palatability to insect larvae. Freed from its role as a food source for plant-eating larvae, its chances for radiation may increase greatly. As a result, the particular substance may come to characterize a genus, a family, or even a larger taxonomic unit. If, however, as does happen, a variety of a species of insect genetically develops the capacity to feed on the "protected" plants, it too may form the basis for a new radiation. This will follow the plant radiation in a coevolutionary process, figure 40.

Mimics, as we have noted earlier, may take advantage of this type of situation. In the Monarch-Viceroy model-mimic system, noted on page 146, the adult, the *imago* stage, of the Monarch has a noxious taste, acquired as the larva fed on milkweed. The larva of the Viceroy, on the other hand, feeds on other plants, willow in particular, and does not impart a noxious taste to the adult. The larvae of the Monarch and Viceroy are very different, showing no mimicry, whereas the imago stages are similar. Coevolution of the Monarch larva and milkweed thus is extended to a more complex relationship involving a mimic-model relationship of the imagos of these two widely separated genera (see plate V).

Hybridization

Males and females of different populations which exist in the same areas, that is, sympatrically, may copulate, and if they are genetically compatible, may produce offspring. This occurs, of course, between

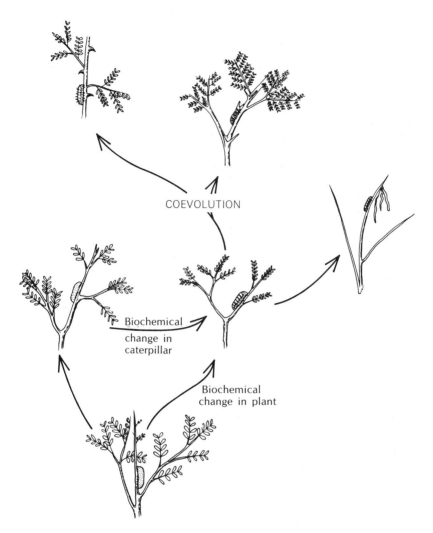

Figure 40 Coevolution. A diagram based on the concept of coevolution of butterflies and plants suggested by Ehrlich and Raven (1964). See the discussion in the text. First, a minor mutation produces a substance in plants which makes them unpalatable to larvae of a butterfly species which has been feeding upon them. The butterflies, freed from this restraint, undergo a mutation which makes it possible for the larvae, once again, to feed in the plants. The larvae then undergo a rapid radiation, adapting to the various lines of plants that have developed. The idea, of course, is oversimplified in the drawing.

races and subspecies, but also takes place between species and genera and occasionally between members of different families. The offspring are *hybrids* and may be totally infertile, or they may be partially or fully fertile with regard to each other or to members of one or both of the parent stocks.

Mules, which are sterile, represent one of the most familiar examples of hybrids. They result from the mating of a male donkey and a mare, usually considered different species of the same genus, *Equus*. Hybridization is not uncommon in nature, but the extent of its evolutionary effectiveness is a matter of debate.

Many plants and animals which normally do not hybridize in nature will do so under artificial conditions when, in one way or another, the natural barriers have been removed. This means, in effect, that the evolutionary aspects of hybridization must be studied under natural conditions, making the problem more difficult than it otherwise would be. Although hybridization occurs in both plants and animals, the different mechanisms of breeding and differences in mobility between the more advanced plants and animals produce different evolutionary results. In plants, a major effect is *introgression* in which the genes of one population become widely dispersed through another population. Introgression is much more limited among animals.

Plants

Of the many well known instances of introgressive hybridization among plants, one analyzed by H. Stutz and L. Thomas will serve well as a case study. It involves hybridization between the cliff rose (*Cowania stansburyana*) and the bitterbush (*Purshia tridentata*). As the names indicate, these plants are assigned to different genera. The morphological differences are extensive (fig. 41).

The two genera have somewhat different distributions, but they overlap geographically in Utah. The cliff rose generally grows at a higher elevation in the area of overlap. Usually, as well, plants of the two species flower at different times, providing a reproductive barrier. On some ridges, however, where conditions of insolation and other factors related to flowering differ strongly from one small area to another, the two flower simultaneously and produce hybrids.

Introgression has developed as a consequence of this hybridization. It has become so prevalent that nonintrogressed populations are rare in Utah. A stabilized hybrid has been established along the southern range of *P. tridentata*, designated as *P. glandulosa*. Such

HYBRIDIZATION

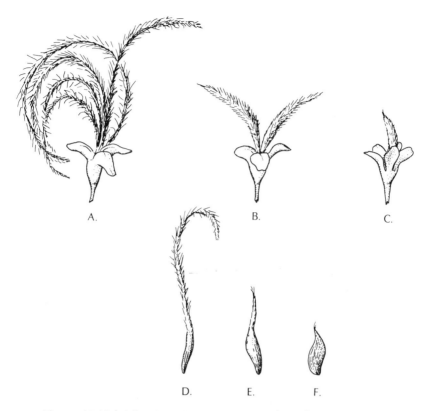

Figure 41 Hybridization. Two genera, *Purshia* (the bitterbush) and *Cowania* (the cliff rose) hybridize under special natural conditions, as discussed in the text. A and C are the fruits of *Cowania* and *Purshia*, respectively. B is the fruit of a hybrid. D is the fruit of *Cowania stansburyana*, E is that of *Purshia glandulosa,* and F the fruit of *Purshia tridentata*. *P. glandulosa* is a stabilized species resulting from a hybridization between *P. tridentata* and *C. stansburyana*. Figures based on photographs in H.C. Stutz and L.K. Thomas (1964).

events are common among plants and clearly play an important role in the establishment of hybrid races, subspecies, and species.

Animals

Hybridization is a common phenomenon among animals, especially where allopatric sister species have developed a zone of contact or overlap. It has been described for a large number of species pairs, with birds and insects comprising the great majority of documented cases. In spite of this, the extent of introgression is usually much less than in plants. Hybrid zones between animal populations appear mostly along narrow zones, being more widespread only if the parent populations have attained a wide common range.

Two species of the common sulfur butterfly *Colias, C. eurytheme* and *C. philodice,* occur together over much of the United States and Canada. The latter species was originally restricted to the northeastern part of the range and did not overlap the more western *C. eurytheme.* With the spread of agriculture, especially of alfalfa, the eastern species (*C. philodice*) has moved west into areas formerly occupied by *C. eurytheme* alone. The two species are interfertile, and the hybrids are fertile and able to mate with each other and with members of the two parent species. The hybrids occur in all mixed populations. Altogether, because females are dimorphic, populations contain nine phenotypes. Because of close resemblances between many of the phenotypes, as well as variations due to slight genetic differences, temperature, and food plants, differentiation is difficult.

In most populations, the percentage of hybrids has stabilized at less than 10%, even where the process has been going on for nearly 100 years. Early studies, based on small samples, indicated that mating was random between the various male and female phenotypes. The low percentage of hybrids, where a much higher percentage might be expected with random mating, was explained by inferring reduced vigor among the hybrids. The only direct evidence, however, was a slight decrease in egg viability.

Recently, O.R. Taylor, Jr. undertook the very complex field and laboratory studies necessary to test this conclusion. Using large samples, he found that mating was not in fact random, but that *conspecific* mating occurs at a higher rate than *interspecific,* or hybrid, mating. To some extent, this is density dependent, with higher interspecific mating frequency in high density populations. It has also been found that hybrids mate preferentially with other hybrids and with *C. philodice,* producing a situation for introgression of *C. eurytheme* alleles into *C. philodice* populations. Under these circumstances, hybridization is restricted, and even over a considerable period of time has not formed distinctive introgressed populations. This situation is in sharp contrast to the case of *Cowania* and *Purshia,* and is more or less typical of the differences of the effects of hybridization in plants and in animals.

C.L. Remington has made a sweeping study of hybridization of species between major biotic regions in the United States. The zones of contact between them, which he called *suture zones*, show extensive hybridization, figure 42. The hybridization involves a wide variety of species pairs, including birds, mammals, lizards, snakes, toads, insects and various plants. Between 20 and 30 pairs are found in most zones with an equally large array of additional suspected hybridizing pairs.

These zones were formed relatively recently, probably as the earlier barriers were removed by climatic changes following the ice ages and, in some instances, by the activities of man. The hybrids do not show superior hybrid vigor or tendencies to spread from the zones. Presumably, the extent of the hybridization results largely from the recentness of the development of the suture zones. Most of the paired species probably were not separated for a very long period, for climatic fluctuations have been rapid relative to usual rates of evolutionary change. Hybridization apparently occurs characteristically between sister species when they first make contact along such zones. Antihybridization mechanisms were not developed while the species were allopatric. They appear to develop rapidly as such things as characteristic odors, calls, visible modifications, and times of breeding arise to channel species into single lines of intraspecies breeding. Extensive as it is, this type of hybridization in animals seems to have little permanent evolutionary effect.

Regressive evolution

Regressive evolution, sometimes called *degenerative* evolution, is a common phenomenon. During evolution, of course, many structures have been reduced or lost in most lines, sometimes remaining as vestiges, but this phenomenon is not what is generally considered regressive evolution. Regressive evolution involves the loss of body systems or their reduction, so that the resulting organisms are much more simply organized than were their ancestors. Much evolution tends toward increasing total complexity, even though some parts of the organism may be simplified, and this increase is often taken as a measure of evolutionary progress. Offhand, regressive evolution seems to be the reverse. If considered alone, with regard only to the forms that show regression, this is true.

Various vertebrates and invertebrates have moved into physical circumstances in which some of their systems are nonfunctional. This is particularly evident in animals that have come to live in cave habitats where pigmentation and the sense of sight have been reduced or

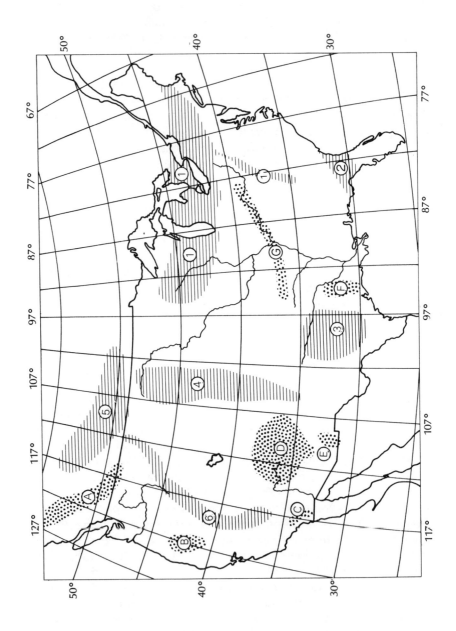

lost. Eyes, for example, may become vestigial. Such cases, as far as the system involved is concerned, do in fact reverse the trend toward increase in complexity.

Many cases of regression, however, occur where there is some sort of *social* organization, either within or between species. In the latter case, this is often called symbiosis, with social systems restricted to single species. In such instances, the system as a whole, the social or symbiotic system, becomes more complex, although its components may undergo marked regression.

Many internal parasites, which with their hosts form elements of a system, are degenerate, having lost those parts of their bodies not necessary to their parasitic existence. Tapeworms, for example, lack a digestive system, which was presumably present in their ancestry, as well as most elements of the sensory system. Many internal parasites lack eyes, even though their relationships indicate that these structures were present in the ancestral stocks. Useless or deleterious organs and systems thus tend to be lost in adaptation to special circumstances of existence. Similarly, of course, new structures may develop in accomodation to new needs, structures for ingestion of food, for maintenance of position, and so forth. Commonly, regression is accompanied by some new or *progressive* features. External parasites much less commonly undergo regression, but tend to be highly adapted to their special roles by modifications of the existing systems.

Changes of the sort just described have at least a partial basis in genetic structure, and selection appears to have been active in both the regressive and progressive stages of development of related species. A different pattern, however, emerges in the development of social structures within a single species. In some types of systems, as among many birds and mammals, roles in social systems are not matched by well-defined morphological characteristics, except as these roles may be related to sex roles. In some insects, however — bees, ants, wasps, and termites — the functions necessary to the societies are distributed among castes with particular social roles and often with distinctive morphology. These tend less to be genetically determined within the species. The next chapter is devoted to a full consideration of the nature and problems of social organization, with emphasis upon the insects.

Figure 42 (*opposite page*) Hybridization in *suture zones*. Hybrids are formed along these zones by species and genera which have very recently come into close contact as climates and plant distributions have changed. See the text for discussion. The map is redrawn from C.L. Remington (1968) by permission of Plenum Publishing Corporation.

Summary

We have examined rather briefly, using mostly particular cases, some of the phenomena of evolutionary processes which go to make up the total of evolution. All of these are to some extent observable in contemporary living systems. They can be studied in life and to some extent by experimentation. It is possible in some instances to relate genotypes and phenotypes directly. The physiological aspects of some of the phenomena can be studied, and the action of the many factors of evolution can be analyzed.

It is through the observations of these and similar phenomena in multitudes of cases and through efforts to understand their biological bases that the fabric of evolutionary processes has been woven. The gathering of data for such analyses is usually an immensely difficult and time-consuming chore. Great care must be taken at all stages to assure that the questions to which answers are sought can properly be asked of the samples obtained. Often it appears that rather simple and elementary questions are answered only after years of careful accumulation of data.

In some instances at which we have looked, for example, the cases of parallelism and convergence, extrapolation from detailed analyses to phenomena at broader levels is possible. For the most part, however, the minor events which reveal the processes of evolution most directly are cloaked as we go to the fossil record and introduce the critical factor of extensive time. A rather large inferential step is required to relate two sets of information and to assess the significance of the various processes and events in a total picture of evolution that has led to our modern diversity.

Table [opposite page] The families of terrestrial mammals from North and South America. The larger sample from South America has 41 families, and the smaller from North America, 26. Nineteen families are common to the areas. The similarity index is 46.

SUMMARY

North America	South America	Common Name
Didelphidae	Didelphidae	opossum
Talpidae	— — —	mole
Soricidae	Soricidae	shrew
— — —	Embollonuridae	bat
— — —	Noctilionidae	bat
Phyllostomatidae	Phyllostomatidae	bat
— — —	Desmontidae	bat
— — —	Natalidae	bat
— — —	Thyropteridae	bat
Vespertillionidae	Vespertillionidae	bat
Molossidae	Molossidae	bat
— — —	Callithericidae	marmoset
— — —	Cebidae	monkey
Hominidae	Hominidae	man
Felidae	Felidae	cats, etc.
Mustelidae	Mustelidae	skunks, weasels, etc.
Canidae	Canidae	dogs, etc.
Procyonidae	Procyonidae	racoons
Ursidae	Ursidae	bears
— — —	Tapiridae	tapir
Tayassuidae	Tayassuidae	peccary
Cervidae	Cervidae	deer
Antilocapridae	— — —	prong buck
Bovidae	— — —	bison, etc.
Dasypodidae	Dasypodidae	armadillo
— — —	Bradypodidae	tree sloth
— — —	Myrmecophagidae	anteater
Aplodontidae	— — —	mountain beaver
Sciuridae	Sciuridae	squirrels
— — —	Octodontidae	degu
— — —	Echimyidae	spiny rats
— — —	Ctenomyidae	tuco tuco
— — —	Abrocomidae	chinchilla rat
— — —	Capromyidae	coypu
— — —	Chinchillidae	chinchilla
— — —	Dasyproctidae	agouti
— — —	Dinomyidae	paca
— — —	Caviidae	guinea pig
— — —	Hydrochoeridae	capybara
Erithizontidae	Erithizontidae	porcupine
Cricetidae	Cricetidae	rats, mice
Zapodidae	— — —	jumping mice
Geomyidae	— — —	pocket gophers
Heteromyidae	Heteromyidae	kangaroo rats
Castoridae	— — —	beaver
Ochotonidae	— — —	pikas
Leporidae	Leporidae	rabbits, hares

IMPORTANT CONCEPTS

ADAPTIVE RADIATION: the exploration of numerous adaptive potentialities of a single genetic and morphological "ground plan." Diversification within a single taxon to fill numerous environmental niches.

ANALOGY: the similarity in appearance of particular structures in different taxonomic groups that results from a similarity of function. A common heritage of the structures is not essential. (Insect, bird and bat wings)

COEVOLUTION: the phenomenon of interdependency of two or more species in time, such that changes in one necessitate changes in the other. SOCIAL: within a species, the development of interdependent, morphologically distinct castes.

CONVERGENCE: the physical or behavioral similarity of two or more remotely related organisms as a result of their similar ways of life. (Fish and whales)

DIVERGENCE: the splitting of a discrete taxonomic group into two or more separate groups evolving in separate directions. The increasing genetic and morphological separation of two or more lines derived from a common ancestral base.

HOMOLOGY: the condition pertaining when structures in divergent taxonomic groups were derived from a single structure in their common ancestor. The derived structures may or may not have similar appearances.

HYBRIDIZATION: the interbreeding, with production of viable but not necessarily fertile offspring, of individuals from two distinct taxonomic groups, usually different species or genera.

INTROGRESSION: the phenomenon of wide dispersal of the genes of one species through the gene pool of another species as a result of hybridization.

MIMICRY, BATESIAN: a system in which a poisonous or distasteful *model* is imitated in appearance or behavior by a harmless species, the *mimic*.

MIMICRY, MÜLLERIAN: a system in which several distasteful species come to resemble one another; a predator that learns to avoid one of the species will subsequently avoid all the members of the complex.

PARALLELISM: the phenomenon of closely related species or taxa undergoing similar morphological changes at about the same time, because both are adapting to similar but noncontiguous environments.

PHYLETIC SPECIATION: speciation in which time is the isolating mechanism. The gradual genetic and morphological transformation of a species through time. The boundary between successive species is, of course, arbitrary.

POLYMORPHISM: the situation in which several distinct and nonintergrading phenotypes occur in interbreeding populations. (*Dimorphism* is the existence of only two distinct forms.)

IMPORTANT CONCEPTS 169

PROTECTIVE FORM or COLORATION: the close morphological resemblance of an organism to some aspect of its surroundings (also called *Cryptic coloration or form*). Also, the existence of color patterns that are frightening or otherwise deterrent to potential predators.

REGRESSIVE EVOLUTION: the gradual loss of a major structure in a particular taxonomic group. (Loss of hindlimbs in whales, loss of eyes in cave fishes.) Usually refers to relatively complex structures.

SUTURE ZONE: a broad area of overlap between adjacent biotic regions, in which hybridization is a common phenomenon at many levels within the two communities. (Plants, insects, birds, etc.)

SYMBIOSIS: close interrelationship between two or more organisms for part or all of their life cycles. There are three types:
Parasitism — one organism benefits, one is harmed.
Mutualism — both organisms benefit.
Commensalism — one organism benefits, the other is indifferent.

REFERENCES

Ehrlich, P., and Raven, P. 1964. Butterflies and plants; a study in coevolution. *Evolution* 18: 586–608.

Kettlewell, H.B. 1961. The phenomenon of industrial melanism in Lepidoptera. *Annual review of entomology* 6: 245–626.

Grant, K.A., and Grant, V. 1967. Effects of hummingbird migration on plant speciation in the California flora. *Evolution* 21: 457–65.

Lack, D. 1947. *Darwin's finches.* Cambridge: Cambridge University Press.

Leppik, E.E. 1957. Evolutionary relationships between entomophilous plants and autophilous insects. *Evolution* 11: 466–81.

Ramirez, W. 1970. Host specificity of fig wasps (Agaonidae). *Evolution* 24: 680–91.

Remington, C.A. 1968. Suture-zones of hybrid interaction between recently-joined biotas. *Evolutionary biology* 2: 321–428.

Stutz, H.C., and Thomas, L.K. 1964. Hybridization and introgression in *Cowania* and *Purshia*. *Evolution* 18: 183–95.

Taylor, O.R., Jr. 1972. Random vs. non-random mating in the sulfur butterflies, *Colias eurytheme* and *Colias philodice* (Lepidoptera: Pierridae). *Evolution* 26: 344–56.

Turner, J.R.G. 1971. Two thousand generations of hybridization in a *Heliconius* butterfly. *Evolution* 25: 471–82.

Wickler, W. 1968. *Mimicry in plants and animals.* New York: McGraw-Hill, World University Library. pp. 88–89.

Watt, W.B. 1968. Adaptive significance of pigment polymorphism in *Colias* butterflies. I. Variation in melanin pigment in relation to thermoregulation. *Evolution* 22: 437–58.

8 The evolution of social systems

Perspectives

Social systems have developed among many different groups of animals in a number of different ways. The one we know best, of course, is our own; and most studies of other systems are based on comparisons to our own societies. This inevitably gives a somewhat biased outlook, but one which is hard to avoid.

Being both social and curious, man has paid a great deal of attention to how his social structures have evolved and where they may lead in the future. Within our own genus *Homo* and, to a much lesser extent, among the higher primates as a whole, social development has depended upon learned behavior. All animals have some capacity to learn, but, compared to man's ability, it is rather rudimentary, even in the most advanced apes. Man, in addition, is able to transfer immediate experiences to broader and more remote circumstances and to transmit concepts and information among individuals through the use of symbolic oral communication. All of man's abilities add up to the possibility of development of an extremely flexible social structure based on learning and to the capability to meet changing circumstances by cultural evolution rather than slower biological evolution.

Being conscious of himself and his future, man can mold his social patterns to his own ends. This ability is in marked contrast with the highly evolved social systems among insects where rigidity is the rule and, to a lesser degree, with the more flexible structures among vertebrates where learning plays a relatively minor role. Not all of man's social activities, of course, are strictly cultural, for ultimately they depend upon his biological nature. Such things as the length of the period of gestation and the need for prolonged child care, for example, make a division of labor and cooperation of individuals a necessity. All primates, as well as many other groups of animals, have these same general attributes. In each, however, they are differently developed, and the social organizations reflect these differences. In each group, biological evolution has furnished the basic characteristics of the organisms, and these characteristics limit the extent and direction of social development on the one hand and, on the other, establish the basis from which innovations may arise.

Advanced social systems have developed primarily among birds, mammals, and insects. Some degree of social organization, however, is found widespread throughout the animal kingdom. Many groups remain virtually unstudied, so that few generalizations over the full kingdom are possible. The line between social and nonsocial organization is fuzzy at lower levels of organization, and only rather precise definitions, which may be too exclusive, can accomplish some degree of clarity.

Our emphasis will be on advanced systems, primarily those of insects and mammals. We will treat the general concepts of social organization, the general nature of mammalian systems, and give special attention to the insects. In subsequent chapters, the social and cultural aspects of hominids will be viewed in the context of primates and man. The study of social insects has a twofold purpose. First, their social systems are intrinsically interesting as a particular result of evolutionary processes. Second, they provide a sharp contrast with the somewhat comparable systems developed among mammals, man in particular.

Social systems in general

A social system can be very broadly defined as a group of individuals of the same species organized in such a manner that there is cooperation within the group. This is approximately the definition given by E.O. Wilson (1971) in his book *Insect societies*. The two critical items are that all members of the system belong to the same species and

SOCIAL SYSTEMS IN GENERAL

that they have cooperative interaction. A broad definition such as this includes a wide range of organizational levels. A classification is necessary to sort out the various levels, and many such classifications have been developed both for vertebrates and insects. Two fairly standard systems are illustrated in the following table.

Vertebrates	Insects*
1. *Asocial*—solitary individuals.	1. *Asocial*—solitary individuals.
2. *Aggregates*—flocks (birds, etc.), temporary association.	2. *Subsocial*—adults care for nymphs.
3. *Reproductive*—short duration, only during reproduction. Division of labor during association only.	3. *Communal*—members of same generation use same nests. No cooperation in brood care by nest mates.
4. *Simple*—persistent, with social hierarchies; no division of labor.	4. *Quasisocial*—adults of one generation use same nest, cooperate in caring for young.
5. *Complex*—persistent, with division of labor; requires communication; both sexes present persistently.	5. *Semisocial*—reproductive division of labor, with worker castes caring for young.
	6. *Eusocial*—same as semisocial, but with overlap of generations so that adult offspring (one or more generations) assist parents in care of parents' young.

*Based on a classification by C. Michener (1969).

The two classifications as listed are similar in many respects. Their differences reflect the basic distinctions between vertebrate and invertebrate societies. There are common features such as aggregation, permanancy, division of labor, and parental care of the young. Among insect societies, emphasis is placed upon the extent and kind of parental care of the young. Castes are developed, whereas they are not in mammals. In vertebrate systems, however, social hierarchies and dominance relationships of individuals are prominent. In both groups, where there is division of labor, behavioral differences occur between those performing different roles, but with insects these differences are in greater or lesser degree matched by morphological differences. Except for sex differences, this is not the case for vertebrate systems.

Social systems of the higher orders (from about 3 on in both lists) are developed only among fairly complex organisms with highly developed nervous systems. Major differences between social systems in vertebrates and insects are related to the nature of the central nervous system. The *brain* in insects has only a relatively few neurons compared to even the simplest vertebrate brain. In addition, the construction and interrelationships of the neurons differ, so that the insect brain is less effective in collection, integration, and transmission of information. Learning among insects is limited and of short duration. What is learned in one situation does not appear to be transferable to other circumstances. In vertebrates, even with the great variation from the simplest to most complex, learning capacity is greater and transference higher, being very high in advanced mammals. Both insects and mammals exhibit innate or instinctive behavior. This trait appears to be more nearly equivalent in them than the capacity to learn and apply the information gained.

These neurological factors, along with behavioral traits consequent upon morphology and physiology, produce the major differences in social systems of the two groups. One of the most definitive and striking differences lies in the emphasis upon modes of communication. Vertebrates rely in large part on the several senses as stimulated by physical actions of others of their species. Chemical communication, while utilized, is not generally of great importance. Insects, while relying on sensory communication to some extent, have a highly developed system of chemical communication depending upon *pheromones* (sometimes called *exocrines,* or external hormones) formed by special glands in various organs. These chemical substances are of great importance in all phases of social life among advanced insect societies.

Insects

Social groups

Well-developed social systems occur only among the Hymenoptera —wasps, bees, and ants—and among termites. In these two widely separated taxa, the social structures are similar in most respects, although there are some basic differences. Of the seven suborders of

INSECTS

Hymenoptera, four have developed *eusocial* systems. These include wasps, with all but one of the social species in a single family among many; ants with many families of the eusocial types; and bees, among which sociality is wide-spread. In each of the groups, a variety of social levels is found. Termites, which are closely related to cockroaches, are all eusocial. There are seven families, and these display different grades of social development within this general level of organization. In the most advanced, Termitidae, caste differentiation is at a very high level.

General features of castes

Each of the social insect groups has similar castes, but these vary widely in details and in their performance of their special duties. Basic to the caste system is the division of labor, and one measure of the extent of social development is how far this division of labor has been carried, both in behavior and morphological modifications related to caste functions.

Reproduction is carried out by one or a few queens in a colony, her prime function being the production of eggs. Colonies are fundamentally organized to assure, as far as possible, the sanctity of the queen and her function. The degree of distinctiveness of castes differs markedly both within and between major groups of social insects. Strictly, it is possible to be accurate only by considering each group separately. In general, however, each group has, in addition to the queen, a soldier caste and a worker caste. Drone males occur among the Hymenoptera, but these are not a part of the social structure among termites. We will briefly summarize the social structures of the major groups of insects.

Wasps   Eusocial structures occur in the family Vespiidae and are highly developed in the hornets and yellow jackets (Vespinae). In the latter, the queen, workers, and males are distinguishable both morphologically and behaviorally (Fig. 43). The nest forms are moderately complex. Within the wasps as a whole, all levels of social complexity and castes occur. At an elementary level, only behavior distinguishes the queen and workers. At no level does a high degree of division of labor occur among the workers.

Bees   The high level of social organization of various groups of bees is well known. Eusocial behavior has appeared at least eight times, and lower levels have been attained over and over again.

Figure 43 Castes of a wasp (*Vespa maculata*) showing only moderate morphological differences among different caste types, A, queen; B, drone; C, worker. Based on drawings by J.B. Betz (1932).

Bumblebees and honeybees show high levels of sociality, whereas others, such as the sweat bees, reach only a low level, eusocial-condition.

Among honeybees the queen and workers have marked morphological, behavioral, and physiological differences (fig. 44). These arise early in the larval stages under pheromone control. The queen inhibits the development of other queens by producing a *queen substance*; but if this restraint is lifted by, say, death of the queen, workers construct queen cells in which they deposit *royal jelly* which stimulates the development of reproductive females. The males in these colonies are drones.

Colony organization is complex. The form of the nest (fig. 45A) and the actions of the workers serve to maintain suitable living conditions, keeping temperatures and humidity at tolerable levels by a

INSECTS 177

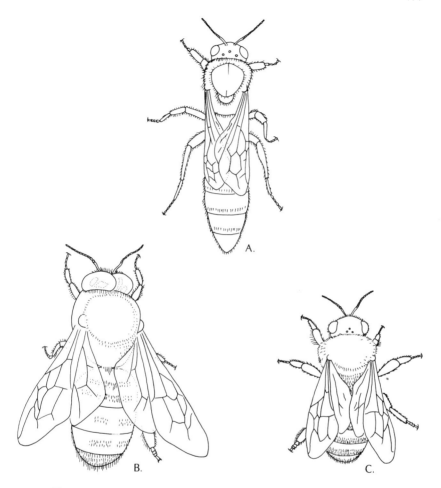

Figure 44 Castes of the honeybee (*Apis mellifera*). Castes are quite distinctive. A, queen; B, male drone; C, worker. Based on photographs in A. Dines (1968).

variety of kinds of behavior. The famous *waggle dance* by which a returning worker indicates the source of food is indicative of the level and intricacy of communication of which honeybees are capable.

Ants In many respects, ants are the most successful of the social Hymenoptera. They are widespread geographically and occupy a great number of diverse environments. They have exploited numerous food sources. It has been estimated that there may be as many as

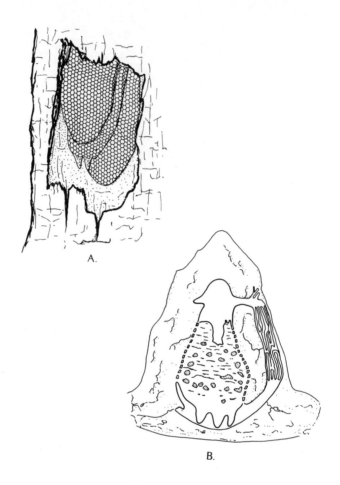

Figure 45 Nests of insects. A, A typical hornets' nest (*Vespa maculata*). B, a complex termites' nest showing fungus garden (center), surrounding air spaces, a vent system for air and humidity control (to the right), and the position of royal cells at the base of the garden. A based on photograph in B. Betz (1932). B based on figure in P.E. Howse (1970).

14,000 species of ants, of which some 7600 have been described. Among these species, of course, there is great diversification in colonial organization, adapted to particular ecological roles. Feeding, for example, may be highly specialized, with one particular food source exploited. Some colonies feed on one special kind of arthro-

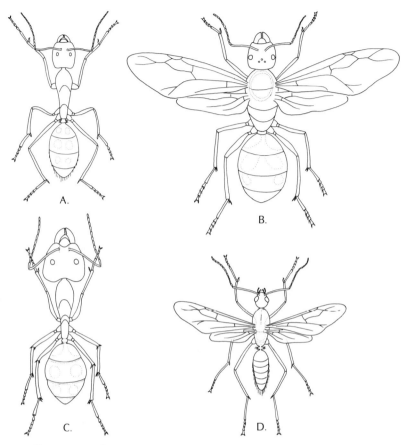

Figure 46 The castes of ants in the spotted sugar ant (*Camponotus maculatus*). A, worker; B, queen; C, soldier; D, winged male. The castes are very distinctive in morphology. Redrawn after S.H. Skaife (1961).

pod, others grow and harvest fungus, while still others feed on exudations of other insects with which they develop complex symbioses.

Some groups of ants construct complex nest types, whereas others do not utilize nests. Various devices of assuring the safety of the queen have developed, and a main thrust of the colony is always in this direction, regardless of the particular way in which it is done.

Basically, ants have three castes,—the queen, worker, and soldier—plus the male, a loose caste (fig. 46). Workers and soldiers are sterile females. Their morphology, behavior, and roles in the social system vary widely in the different groups of ants. Ants have extended their colony structure by embarking upon slavery and social parasitism. In some instances, this condition has gone so far that the enslaving ants have become dependent on the slave groups, acting as parasites upon them.

Termites The social structure of termites is highly developed and differs in some ways from that of the Hymenoptera. Castes are similar, with workers, soldiers, and primary reproductives, the queens. Both sexes, however, are workers and there are no male drones. In addition, there are larval or nymph castes, immature individuals that participate in colony work. Within the termites' very complex castes, hierarchies have developed and the behavioral and morphological distinctions of castes are often great.

Colony activities are highly complex and tend to be distinctive of species. Nests are not only complex and highly efficient but are species-distinctive and show evolutionary series. The complex structures are met by complex behavioral patterns by which the workers maintain nest conditions, such as temperature and humidity, and soldiers defend the nests, taking advantage of their own structure and mobility and the configuration of nest structure (fig. 45B).

Origin of castes

The origin of castes is one of the most challenging and fascinating areas of the study of development and evolution. Within social systems, castes, except those directly involved in reproduction, perform tasks that benefit the queen and the colony but are likely, at least part of the time, to be disadvantageous to members of the castes themselves. The workers tend the young, serve the queen, and aid in keeping the colony liveable. Soldiers defend the colony, although they may be destroyed in so doing. Where drones are present, they serve only to fertilize the queen. These acts, which benefit the colony but are disadvantageous, or at least not beneficial, to the performers of them, are usually called *altruistic*. At once, the question arises as to how natural selection, which alters the genotypes in successive populations by affecting individuals, can operate to bring about a caste composed of individuals that do not contribute their genes to the succeeding generations. This question has not as yet been answered to the satisfaction of everyone, but later in this chapter we will look at some of the suggestions as to how this may take place.

At a somewhat different level, the origin of castes within a single colony is of considerable interest. Arguments as to whether castes at

INSECTS

this level are or are not genetically determined have gone on for a long time. The question can be stated in another way: Do all eggs have an equipotential of developing into any one of the caste types, or is this already determined in their genetic composition, that is, in differences between eggs which develop into one caste or another? Among the Hymenoptera one major difference between eggs exists. Some are fertile and some are infertile. The latter produce males by *parthenogenesis*. Because there is no fertilization, no pairing of the chromosomes from a male and female gamete, these eggs are haploid. Haploidy may have very important consequences in the evolutionary origin of the colonies, but strictly is not a basic genetic difference between eggs of a colony.

It now seems quite clear that, among the hymenopterans, castes are determined nutritionally and not by genetic predisposition of the eggs. There is one likely exception among the stingless bees. Even if a basic nutritional cause is generally applicable, the process of caste determination is far from simple. In ants, for example, at least six interacting factors, including, among others, an interplay of nutrition and periods of hibernation, are involved. Equivalent but different complex controls exist in both bee and wasp colonies, with the latter being somewhat simpler in general.

Termites, although they develop similar castes, show rather different patterns. Growth and development is regulated in termites, as throughout the insects, by true *hormones*, substances secreted by the endocrine glands. In some of the less complex termite societies, it has been demonstrated that pheromones supplied by developing larvae alter endocrine responses and channel development into different avenues, depending upon the time, circumstances, and amounts of the interrupting pheromones. Thus, caste development, rather than being primarily controlled by nutrition, is largely pheromone-controlled. In more complex systems of social organization among termites, the precise nature of controls has not been thoroughly determined, although some of the factors have been isolated.

Caste development within colonies is primarily under the direction of events within the colonies, and the queen tends to occupy a dominant position in guiding this development. From simple to complex colonies, however, changes and refinements of caste-producing mechanisms and caste behavior occur in a fairly regular fashion, one that appears to be phylogenetic. Worker castes in termites, for example, show increasing reduction of some systems—reproductive structures, the gonads, and the eyes, for example. These changes are clearly genetic in origin, as are various other physical and behavioral characteristics related to the particular pattern of caste development. Thus, although within colonies the production of castes is not genetic, the capacity to develop castes and the particular course of development have a genetic base and evolve.

The evolution of social systems

General nature and problems

The problems of the origin of social organization by natural selection was treated by Darwin in his *Origin of species* and has been of concern to many students for the last century. In all systems of moderate complexity, some individuals make sacrifices which benefit others and benefit the group to which they belong as a whole. Behavior of this sort is generally recognized as altruistic when the donors lessen their own chances for success. In terms of *fitness,* referring to the capacity to reproduce, altruists perform their tasks in such a way that their fitness is reduced.

Altruism is characteristic within family units, those in which there is a strong genetic relationship between the individuals. For the most part, members of populations that lack close genetic relationships are competitive. There are, of course, exceptions, but it seems safe to assume that altruism is, in the final analysis, genetically controlled.

Given this assumption, essentially that such behavior is genetically determined, the basic evolutionary problem can be formulated as follows: How can natural selection favor genes that result in self-sacrificing behavior which benefits the genetic competitors of their bearers? In one way or another, this question has been raised by many students. The statement given above closely follows the formulation of Williams (1966).

Altruism by parents is a necessary condition of evolutionary success of organisms which engage in significant care of young. The selective success that it assures overrides the cost of altruism inherent in some loss of reproductive capacity. Among vertebrates, this is a common phenomenon within birds and mammals, but is also found widespread through other classes.

Extension of altruism beyond the small family, in which it is exercised by the parents alone, poses somewhat different problems. In larger assemblages, altruistic acts may be conferred on members of the species that differ more from their donors than is the case in the limited family group. This has led to various concepts of selection at the superfamilial level, that is, colony selection. The general term applied at such levels is *group selection*. Darwin's considerations led him in this direction. Later students have developed more sophisticated models which include genetics, of which Darwin was unaware.

The general idea of such models is that the loss of fitness of some members of a society is compensated by reproductive success of genetically related phenotypes, so that the colony as a whole is favored. Because, however, the altruistic individual does not provide his own genotype to the succeeding generation, it would appear that acts of altruism would be statistically negative with regards to his fitness, and to the fitness of the group or colony as a whole. Hence, an argument about group selection has developed and still flourishes.

Social groups in various animals, including *Homo sapiens* in primitive human societies, may include members that are not genetically closely related. Factors other than those active in populations of genetically closely related individuals are present, and these will be touched upon in subsequent chapters. For the remainder of this chapter, attention will be focused upon eusocial systems among insects. Members of these systems are genetically very similar, and many of the problems of social evolution come into focus in view of the highly complex but precise nature of many insect societies.

Social insects

The origin of insect societies

Eusocial systems, as noted earlier, are present among insects only in the Hymenoptera and termites. These form but a limited part of the total array of types of insects. In both the hymenopterans and termites, systems of organization range from very simple to highly complex. Eusocial systems in both include a high percentage of permanently sterile offspring and successive overlapping broods in the colonies that have common parents. Thus, brood management by siblings is devoted to genetically similar individuals. If the chances of success of the reproductives, two individuals in advanced social organization, were enhanced substantially by colony activities, then the cost of infertility of most colony members could be offset. If this enhancement were of major proportions, then the gene or genes basic to altruistic behavior of the nonfertile colony members could be in a favorable selective position under the usual processes of natural selection.

Many complications arise, different for different types of colonies, in the application of this general model. Basically, however, it appears to provide a valid interpretation of the way that altruism can arise through natural selection. The concept has come largely from the work of S. D. Hamilton, summarized in Wilson (1971).

As we have seen earlier, most evolutionary events that have led to exploitation of some particular way of life in a new habitat zone have arisen as the result of a key innovation. Such a change only makes a new direction possible; it does not assure it. Eusocial systems, as well as their less complex precursors, probably had such an origin. It appears, however, that different key factors were operating in the cases of the hymenopterans and the termites, so that they must be treated separately.

Hymenoptera

Complex social systems have arisen at least eleven times among the hymenopterans in wasps, bees, and ants. Among all other insects, except the termites, such systems have not developed. Although different hymenopterans have different social organizations, the basic resemblances are sufficient that they probably have resulted from some common property, a property that presumably is not widespread among the insects as a whole.

One such peculiarity of the hymenopterans is that the males are produced from infertile eggs, that is, they are parthenogenic. As a result, they are haploid, having but one set of chromosomes. The females, on the other hand, are produced by fertile eggs, and, of course, are diploid. The phenomenon is termed *haplodiploidy*.

When the female (queen) is fertilized by a haploid male, all daughters receive the same set of genes from the father, but each receives one or the other of the two sets from the mother. In the course of his extensive work on the origins of eusocial systems, S. D. Hamilton developed a model of how haplodiploidy may result in the development of eusocial systems. A simple explanation of this has been given by E. O. Wilson in his book *Insect societies* and what follows has been paraphrased from his account.

It follows from the axioms of population genetics that an altruistic trait can develop if the sacrifice of fitness of the altruistic individuals is compensated for by increased fitness of some group of closely related relatives. Fitness, as before, refers to the capacity to produce offspring. For altruism to develop, the fitness factor of the relatives must be greater than the reciprocal of the *coefficient of relationship* of the altruistic members and that of the relative group. This coefficient is the equivalent of the average fraction of genes shared by common descent. In usual diploid systems, this is ½ for sisters, ¼ for half sisters, and so on.

SOCIAL INSECTS

It follows that if some individuals make reproductive altruistic sacrifices, then the role of the sisters must be more than doubled, that of half sisters quadrupled, and so on. The full effect of fitness in the total array weighted in this fashion is termed *inclusive fitness*. It is evident from this, among other things, that closeness of relationship may be a most important factor in the development of altruism by natural selection.

The existence of haplodiploidy results in a change of these ratios. The coefficient of relationship between sisters becomes ¾ rather than ½. In highly organized colonies, female offspring are the product of a single male and female, that is, they are full sisters. Thus, the ¾ ratio holds, whereas between mothers and daughters it is still ½. It is thus advantageous, relative to the development of altruism, that there is a sister-sister rather than mother-daughter care. This will be positively selected for. The ratio in sisters results, of course, from the fact that the females receive the same genes from the male parent and on the average, half the same from the female, which may be expressed as follows:

$$(1 \times 1/2) + (1/2 \times 1/2) = 3/4$$
$$\text{father} \qquad \text{mother}$$

When, as in eusocial systems, the mother lives beyond the time of maturation of her young and care of sisters by sisters of successive generations occurs, this care can increase the inclusive fitness over that given by care of the mother to her own offspring.

Everything else being equal, hymenopterans, being haplodiploid, should tend to become social. The theory is consistent with much of the available evidence. Various objections have been raised. The existence of more than a single queen in some colonies or the possibility that a queen may be fertilized by more than one male pose problems.

Nevertheless, the theory has importance. More study is necessary to document it. If the theory is supported, and currently it seems to be a rational basis for explanation, then it is a major contribution in its own right. If it is found to be false, or only partly applicable, then it will have served, as have most theories of the past, as an important stimulant and focal point for research.

Termites

Termites do not have the property of haplodiploidy, so the explanation advanced for the hymenopterans cannot apply to them. They have, however, as we have seen, developed social systems that range from moderately simple to highly complex. One feature that they possess assures some degree of social behavior and may provide the key innovation in back of the development of their highly organized social structures. They are wood-feeders. The capacity to utilize wood (cellulose) as a food depends upon a symbiotic relationship with intestinal flagellates. These are passed on from older termites to the young by anal feeding. This feeding is a necessity, and the relationships between mature and young cannot be dispensed with if termites are to survive.

Neuter castes of the termites include both males and females and there is no high coefficient of relationship as found among the hymenopterans. Thus, although a social bond may occur by the means suggested, anal feeding and the transmission of intestinal flagellates and the development of higher levels of donorship, including the complex caste systems that have arisen, do not necessarily follow. Although the sibling groups are not as close genetically as in the hymenopterans, they do have the kind of family relationships envisaged in the general model of Hamilton.

Many studies have been made on how termite eusocial systems, in the absence of haplodiploidy, might have arisen. Williams and Williams (1957) developed a model that demonstrated that even though development of social traits, including altruism, or as they called it *donorism*, were deleterious reproductively within sibling groups, these traits could become advantageous by competition between sibling groups of the same species. This represents a development of group theory with selection at the family level. In a later work, G. Williams (1966) discussed the problem further, putting forth a possible alternative explanation related in particular to the intensity, timing, and duration of sterility. Under this concept, adjustment to seasonal and ecological conditions could be such that partial, or temporal, sterility might be advantageous and lead the way, through evolutionary changes, to a high level of development and permanency of sterility as more advanced social systems arose.

As matters stand, a base for the beginning of termite social systems seems established, the initial cooperation attendant upon the transmission of intestinal flagellates. Various suggestions of how higher organization may have come about have been made. Special circumstances clearly *have* favored the development of termite eusociality, but just what these are, how they have operated in different groups of termites, and what the courses of changes have been are not well understood.

Evolution of complex social systems

If we assume now that the bases for explanation of the beginnings of eusocial systems have been developed, or at least that we have a plausible array of explanations, the problems of the full evolution of these levels still remain. Like the origins, the subsequent evolution of the systems remains cloudy. First of all, the systems have developed independently in a fairly large number of lines of descent with extensive parallelism and convergence, and a wide variety of eusocial systems has arisen. They are adapted to various environmental circumstances and are expressed in many life modes. Yet there are underlying resemblances in all systems.

It may be asked, for example, why different social levels exist in closely related stocks. Why have some species of wasps, for example, become highly organized, whereas others have remained extremely simple? Have complex systems in some instances reverted back to less complex ones? Why do some societies have a great number of individuals, whereas others have only a hundred or so? In some the castes are highly polymorphic, in others they are not. Some social units have several queens, others have but one. All of these features, plus others of similar nature, presumably have evolutionary explanations. Most are not well understood, and, at best, we can hope currently to extract some general ideas that are more or less widely applicable. As before, the remarks that follow depend in part on interpretations of the extensive analyses of E. O. Wilson (1972) who has synthesized the work of many students, including his own.

Very generally, the highest levels of organization occur in the largest colonies, although the correlation is rather loose. Large size of populations on the whole has a number of advantages—it provides a defense against many types of predators, it improves homeostasis, it makes for more efficient labor, and it confers a capacity to perform tasks that smaller populations cannot. Construction and operation of complex termite nests or the movements and activities of a full colony of army ants require large numbers.

On the other hand, the rate of productivity of adult females falls off in larger populations, a point stressed in particular by one of the leading students of social insects, C. Michener. Thus, in small populations the proportionate number of sexual forms per generation is greater and the capacity to form new colonies relatively enhanced. Small colonies tend to be less subject to predation by mammalian predators but are less able to defend themselves against various invertebrate predators.

These factors act, on the one hand, toward the evolution of large colonies and, on the other, toward retention of small ones and the reduction of large colonies when formed. Just how the factors operate and how they, and many others like them, balance out depends in large part upon the particular biological properties of the colony and the environment in which it exists and evolves.

As many aspects of colonial organization and size are studied, similar compromises emerge giving advantages to one type of colony under some conditions and to another under different conditions. Workers in large colonies, for example, are in general less productive per individual than those in similar smaller colonies. To some extent, this may offset the advantages of size, but the large colony persists in spite of it, meeting some circumstances that the smaller one cannot.

Size, of course, is but one factor, and it operates somewhat differently in major colonial types. A primary question of another nature, for example, is why colonies tend to evolve to have but a single queen. Not all do this, but the tendency is strong and widespread. Colony size may be a factor, but various other suggestions have been made. In line with the model of Hamilton in which genetic closeness of colony members is a critical advantage, the emergence of a single queen should produce the relatively most selectively advantageous colony. Competition among queens, in contrast to lack of competition among other colony castes, may be an important factor in producing highly developed societies and may in itself be selected for. Workers, however, also have agressive tendencies toward queens, tending to kill them if more than one is present, but always leaving at least one. Is this tendency developed selectively?

If the position is taken, as it is in this chapter and by most students of the subject, that the processes of natural selection as understood are sufficient to account for the evolution of social insects, one point stands out. A great deal of careful and detailed study, based on concepts of population biology, is required to provide a full understanding of the evolution of societal organization. All aspects of colonies must be studied in the field as a basis for construction and testing of models of the sort considered in this chapter. Perhaps, in the course of such studies, new principles not now recognized will emerge. It may be that the theoretical framework of natural selection as now conceived will have to be modified. At present, at least, explanation within the processes that are considered basic seems possible, but we are far from a fully satisfactory understanding of many of the phenomena themselves, let alone a full explanation of their evolution.

IMPORTANT CONCEPTS

ALTRUISM: in general, behavior performed for the benefit of others. In context of social organization, acts which benefit the social group but may be disadvantageous to the fitness of the individual performing the acts.

CASTE: a discrete subdivision of a social group having a specific function within that group. In insects, morphologically distinct groups within the colony, each of which assumes a particular responsibility in the division of labor of the colony.

COLONY: a group of organisms (either animals or plants) of the same species, living in close association with some degree of interdependence among individuals.

DIPLOID: individuals having two complements of similar chromosomes, usually one set from each parent.

DIVISION OF LABOR: the work necessary to sustain a colony is divided up among subgroups of the colony, thus, no individual must perform all life-perpetuating behavior. Specialization, with improved efficiency, results.

DONORISM: in the context of altruism, the term applied to sacrifice of fitness of an individual (the donor) for the benefit of the colony as a whole.

EUSOCIAL SYSTEM: a social system in which adult offspring assist the parent(s) in the care of young produced in subsequent generations.

FITNESS: an expression of the reproductive capacity of an individual or group.

GROUP SELECTION: the condition pertaining when the colony, population, or other type of social group becomes the unit of selection rather than the individual.

HAPLOID: individuals having but a single complement of chromosomes.

HAPLODIPLOIDY: the condition in which one sex has a single complement of chromosomes (haploidy) and the other has a double complement (diploidy).

HOMO: the single genus within the primate family Hominidae. Includes the species to which modern man (*Homo sapiens*) and certain closely related extinct species of man belong.

HOMEOSTASIS: the condition of a system (physiological, ecological, thermodynamic, etc.) in equilibrium; pertaining to a steady-state system capable of maintaining equilibrium in the face of external changes.

HORMONE: a chemical substance secreted into the blood stream by endocrine glands which affects the physiological activity of its target organ or tissue.

PARTHENOGENESIS: the maturation of eggs into normal adults without fertilization.

PHEROMONE: a chemical, gland-produced substance used for communication among organisms of the same species.

SIBLING: a brother or sister.

SOCIAL SYSTEM: a group of conspecific individuals whose behavior is organized so that they can act cooperatively.

REFERENCES

Betz, J.B. 1932. The population of a nest of the hornet *Vespon maculata*. *Quarterly Review of Biology* 8: 197–209.

Dines, A. 1968. *Honeybees from close up*. Photographs by S. Dalton. London: Cassel Ltd. pp. 4–5.

Howse, P.E. 1970. *Termites: a study in social behavior*. London: Hutchinson University Library. p. 107.

Michener, C. D. 1969. Comparative social behavior of bees. *Annual Review of Entomology*. 14: 299–342.

Skaife, S.H. 1961. *The study of ants*. London: Longmass, Green and Co. Ltd. pp. 15–18.

Williams, G. C. 1966 *Adaptation and natural selection*. Princeton: Princeton University Press.

Williams, G. C., and Williams, D. C. 1957. Natural selection of individually harmful social adaptations among sibs with special reference to social insects. *Evolution* 11: 32–39.

Wilson, E. O. 1971. *Insect societies*. Cambridge: Harvard University Press, Belknap Press.

9

The evolution of man

Perspectives

The evolution of man is of special interest not only because our own subspecies *Homo sapiens sapiens* is involved, but because it portrays features not seen in other evolutionary sequences. The most important of these is the development of physical and biological properties which, combined, result in the origin of a highly organized social system. The system depends in large part upon learning and symbolic communication which have led to cultural evolution. The development of cultural evolution has had an immense effect upon the course of Darwinian evolution in man. For these reasons, in this and succeeding chapters special attention will be given to what man is and to the nature of his physical and cultural evolution.

The broad course of man's physical evolution is now fairly clear, but many details remain to be filled in. On the basis of the fossil evidence, his history has followed a course independent of other primates for at least 14 million years. Origin of the hominid stock, the family to which man belongs (fig. 47), took place in the Old World, probably in the tropics but exactly where is uncertain. Initiation of ground feeding and of a terrestrial way of life probably were the most

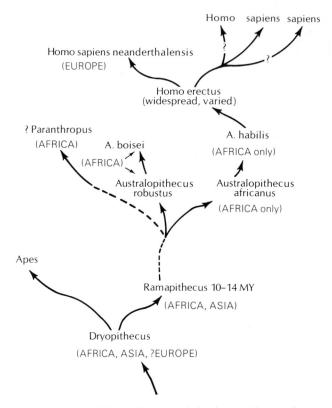

Figure 47 A possible phylogeny of the hominids. As discussed in the text, there is not full agreement on the relationships of members of this family. *Dryopithecus* is a probable ancestor to man, leading to the problematical hominid *Ramapithecus*. From stage two, and possibly three, lines of *Australopithecus* are shown, the more lightly-built forms—*A. africanus* and *A. habilis*—forming one and the heavier—*A. robustus* and *A. boisei*—forming the other. The question of *Paranthropus* as a separate genus is indicated. Most persons consider this genus of heavier forms to best be included in *Australopithecus*. Homo is thought to have come from the *A. africanus*–*A. habilis* line. For the radiation of the races of *Homo sapiens sapiens*, see figure 53.

important initial aspects of hominid evolution. Adaptations to this kind of life were superimposed upon primate features which appear to have had an arboreal background. Among these adaptations, the development of bipedalism was extremely important. Early in their history, members of the hominids became hunters and gatherers, and soon they began to fabricate and use rudimentary tools. Thereafter, the principal physical evolution involved perfection of bipedalism and increase in size and complexity of the brain. These two modifications were accompanied by an accelerating cultural evolution.

That such a reasonably concise statement can be made represents a triumph of major proportions over the two main problems that have hampered the study of the evolution of man. One has been the paucity of fossil remains, which has made the course of the actual history largely speculative. Although the record still is scant, it has become good enough to provide a reasonable basis for interpretation of major events.

The other difficulty is equally important and, if anything, is less easily remedied. Man has been just too close to the problem to evaluate his own evolution objectively. This situation appears at two levels with considerable overlap. First of all, it has affected the scientists who have studied man. Such studies have frequently been treated as something apart from the rest of evolution. With a better fossil record and more dedicated efforts of the students of man to avoid this difficulty, the situation has been greatly improved, although it is still far from perfect.

Secondly, evolution, as applied to man, has come directly into conflict with doctrines and beliefs, mostly religious in nature, which have provided an important fulfillment and source of continuing reinforcement of people's need for dignity and importance. A pertinent example is to be found in the uproar of opposition that arises when the genetic studies of races to determine their inherited differences, are proposed, especially if there is suspicion that downgrading of one or another group might possibly emerge. More general is the religious opposition to the whole idea that man has evolved, rather than that he was created *de novo*, as portrayed in the doctrines of many different faiths.

Yet another aspect is that man, conscious of himself, does not consider his species suitable for controlled experiments of the sorts regularly carried out on other life forms. Opposition, of course, also develops to "perpetrating such indignities" on other species, with the degree of opposition largely relative to the level of organization of such organisms and their social relationships to man.

All of these factors have made the objective study of the nature and evolution of man difficult. Within the limits of available materials his physical evolution has been studied thoroughly. As the physical changes have become better understood, attention has been turned toward studies of the factors underlying the observed events.

What man is physically depends upon natural selection, the directions of change resulting from the effective reproductive capacities of different genotypes. As in other species populations, behavior and social structure are highly significant with relationships between the physical expression of the genotype and the environment.

The special problems which arise in the study of man, beyond the psychological barriers noted above, stem largely from the nature of his social structure and behavioral characteristics. The potentials of behavior are genetically determined, and behavior itself is the continuing adjustment, within the limits of this potential, to the environment. Man's inherited behavioral potential is more complex and less stereotyped than is the case for other species of social organisms. Consequently, responses to similar conditions may be highly varied. Environmental factors thus have come to play complicated roles and are difficult to isolate, study, and relate to genetic heritage.

Over the millions of years of hominid evolution, the social structure of populations and the potentials of the genetically based behavioral patterns have altered and increased greatly in complexity. The environment in which hominids lived has changed with time and has been modified as populations have spread into different parts of the world. Behavioral and physiological flexibility have permitted a tolerance of widely different physical and social environments. Furthermore, hominids have increasingly been able to manipulate their own environment. Social structure has been altered by these capabilities, differently under different circumstances, while itself providing an important selective system in which hominid evolution has taken place. A social flexibility not found in other social organisms is characteristic of man.

This sort of evolution in which evolving social structure is conditioned by the organisms with feed-back to the organisms in a selective way is, of course, not unique to hominids. We saw it in full development among the insects, and it occurs to varying degrees in all social organisms. In hominids, the physical and behavioral adjustments to social roles are less patterned by the inherited and ontogenetic features of the phenotypes than in most other social animals. Furthermore, highly effective ways of transmitting information and experience to succeeding generations have developed during human evolution, making possible the rise of a very complex culture which developed rapidly and took on many different forms.

The evolution of man, then, although not apart from other evolution, has added many aspects that make it more difficult to understand. Partly, this is merely a matter of its complexity. The unique feature of self-consciousness, which itself makes the study of man by man possible, creates psychological barriers to study at the same time. These barriers, which tend to substitute subjectivity for objectivity, make it urgent that man comes to understand his evolution scientifically, so that he may better be able to judge what he is and how his own future, which he can control, may best be directed.

What is man?

If the term man is used to refer only to modern man, that is, to the subspecies *Homo sapiens sapiens,* the answer to the question *What is man?* poses no problems if viewed biologically. The taxonomic position of *H. sapiens sapiens* is shown in figure 48. Man is a vertebrate, an animal with a backbone, and a mammal, having such features as hair and mammary glands. Among mammals, he belongs with the placentals on the basis of his reproductive processes and structures. Continuing to lesser levels, we find that man has many features in common with lemurs, monkeys, and apes and thus is readily placed within the Primates. Apes and hominids form the hominoids, a subgroup within the order Primates.

Hominids are long-legged bipedal hominoid primates. They have relatively large brains, but the brain size in the most primitive representatives is comparable to that among the closely related apes. As man evolved, both the absolute and relative size of the brain increased, a trend but little expressed among the apes.

Enlargement of the brain is moderately easily traced among fossils, and for this reason figures prominently in discussions of the changes leading to *H. sapiens sapiens*. Proportional changes of different parts of the brain also were important but are more difficult to determine from fossil materials. When the brains of living apes and man are compared, major differences are found. The cerebral hemispheres of man are disproportionately large. The number of cells, however, is much less than would have been the case were enlargement merely a matter of addition of cells. Not only are the cells in man larger and more complex, but they are more widely spaced (fig. 49).

Such modifications occurred during the evolution of Homo from more primitive hominids. Because they are not well reflected in the form of crania and cranial casts, little direct information upon the times of origin of different features is available. Inferred behavioral patterns of early fossil hominids, especially their capacity to make and use tools, have suggested that basic organizational changes had occurred prior to the time that brain size increase became important.

THE EVOLUTION OF MAN

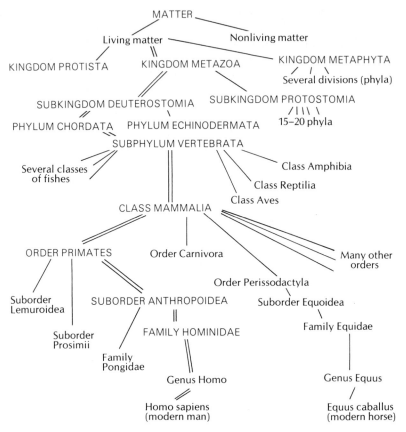

Figure 48 A partial flow-diagram classification of matter and organisms showing the relationship of the primates to other groups of organisms. For simplicity, all unicell organisms are placed in a single group, Protista, which includes both procaryotes, often classed as kingdom Monera, and single-celled eucaryotes. Multicellular plants are Metaphyta and multicelled animals are Metazoa. The double lines show the course to hominids and man. One other order of mammals, the ungulate Perissodactyla (horses, rhinoceroses, and tapirs) is shown in its parallel development. At each stage, the branching off of other groups of animals is shown.

Many anatomical features of hominids relate to their habits of walking erect. The backbone has a somewhat S-shaped curve, the pelvis is rotated and modified in shape to accomodate the femur, the upper limb bone, and muscles of the legs, back, and abdomen. The feet do not have the *handlike* structure found in most primates but

WHAT IS MAN?

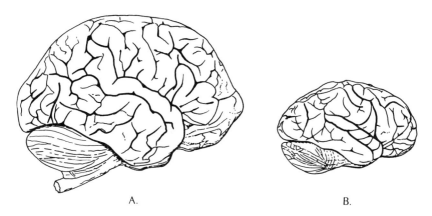

Figure 49 The brains of A, modern man, *Homo spaiens sapiens*, and B, the chimpanzee (*Pan troglodytes*) drawn to scale. Anterior is to the right. Note especially the size difference and the differences in proportions of comparable parts of the brain, with the intervening grooves, or sulci.

have short toes and an unopposed hallux, or *big toe*. The heel tends to be well formed, especially in more advanced hominids.

The forelimb is relatively short in *Homo sapiens sapiens* and has undergone some changes from the conditions in early hominids. Various features of the forelimb indicate that the arms were used for suspension in the ancestors of man, rather than being involved in *knuckle-walking* which is characteristic of gorillas and chimpanzees. Vestiges of the stages in which suspension was a primary function are found in the skeletons of primitive fully bipedal hominids, the australopithecines.

The toothrow in *H. sapiens sapiens* is short and forms a curved arc. The canine teeth are small relative to other teeth and do not interlock. Other teeth as well are proportionately small compared to those of apes and most other hominids. The enamel, however, is thick on the crown surfaces. In apes the molar and premolar teeth, the grinding battery, are disposed in nearly parallel rows. Canine teeth tend to be large, especially in males, and molars and premolars are strongly developed. The less advanced hominids are somewhat intermediate between the condition in apes and man proper, and many of the dental features of the latter have developed within the family Hominidae. Along with modifications of the teeth, the muzzle has become less protuberant and the face more vertical. A distinct chin has developed.

The result of various changes has been a marked increase in the efficiency of jaw action (fig. 50). With shortening of the face, the

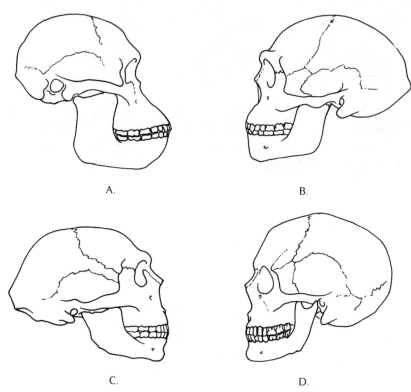

Figure 50 The skulls of four hominids in lateral views. A, Australopithecus; B, Homo erectus; C. Homo sapiens neanderthalensis; D, Homo sapiens sapiens. Not drawn to scale but to more or less common size to indicate resemblances and differences.

external jaw muscles, the masseteric muscles, have come to lie relatively far forward, and the ascending ramus of the jaw has become nearly vertical. Other modifications of muscles and articulation of the upper and lower jaw have made possible more effective lateral motions. The net effect is development of a powerful biting and grinding action, well suited to dealing with harsh foods. Accompanying these changes have been modifications of the crown surfaces of the teeth which have produced a grinding surface with a series of low cusps and the durable caps buffered by thick enamel.

These are some of the physical trends of the hominids which find their full expression in *Homo sapiens sapiens*. They set this group apart from all other primates but, at the same time, closely resemble many features of various apes, both living and extinct. Structures of the skeleton and dentition have been emphasized because they are the physical characteristics which can be studied in fossil remains. Characteristics of the soft anatomy and physiology as well show both resemblances to and differences from those of their counterparts in the apes. The differences appear to be adaptive, being strong where they are involved in distinctly different ways of life and less so where they are not.

Physically, the hominids are about as far from the ancestral lemuroids as are modern horses from their ancestor eohippus (*Hyracotherium*) (fig. 51). At the beginning, as is true for horses and their

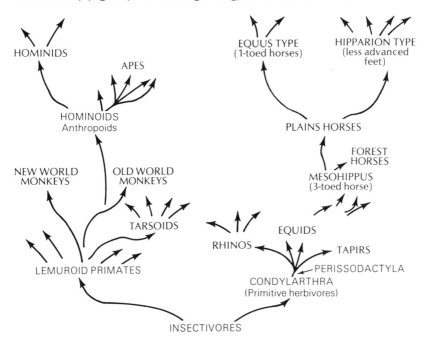

Figure 51 Comparative phylogenies of the order Primates and the order Perissodactyla (horses, rhinoceroses, and tapirs). Note the similar early radiation and then the branching after the *Mesohippus* stage of the horses (Equidae) and the branching of Primates after the origin of manlike forms (Hominidae).

near relatives, the skeletal and dental remains of all primates were much alike. The skeletal and dental remains of the initial hominids and hominidlike apes, known as *fossils*, are so much alike that assignments to one group or another are matters of debate still not entirely settled.

For all the physical similarities between *Homo* and living apes, gorillas and chimpanzees in particular, an immense gap in behavior exists. It is chancy, of course, to say that any feature of modern man, even his ability to communicate by abstract symbols, is unique. Rudimentary development of many of his characteristic features, including this one, have been detected in other organisms. However, the combination of self-consciousness, use of articulated speech and abstract symbols in writing, and the construction and use of complex tools sets *Homo sapiens* well apart from all other animals. This very apartness has been the basis for arguments that modern man could not have arisen through evolution and that his emergence was not through gradual change. If we look only to the modern apes and modern man, this may appear reasonable, for the gap is great. The *unique* features in many instances, however, probably were but slightly manifest in the earliest hominids. It was during the development of man that the special behavioral characteristics of modern man came into being.

Along with the physical and behavioral modifications, both social and cultural organization evolved. As we go back in time, the evidence of these phenomena becomes increasingly difficult to obtain, but almost to the beginning of the hominids, there are some hints. Later in this chapter, we will examine some of these.

To answer the question *What is man?*, that is, *Homo sapiens sapiens*, we may say that he is currently the end member of an evolved sequence of primates called hominids. He is a creature that is highly distinctive physically, behaviorally, and socially. He is different from other hominids, but in degree not kind. The coordination of physical, social, and cultural aspects which developed within the hominids has produced a species very different from any which has gone before, the greatest difference being the development of self-consciousness. Man may thus be defined as the self-conscious animal. The rest of this chapter will be a study of how he arrived at this position.

The fossil record of man and his relatives

Figure 47 gives a brief summary of the phylogeny of hominid primates related to man's evolution based upon the available fossil evidence. Although this is a fair statement of what is now known, it must be

THE FOSSIL RECORD OF MAN AND HIS RELATIVES

recognized that many uncertainties exist and that changes will be made in the future. Very recently (not formally published at the time of this writing), a find of a *large-brained* hominid of about 2.7 MY in age has been reported by Richard Leaky. This comes from the Lake Rudolph region of East Africa near where very old tools have been found. Definitive information is not yet available, but it seems that this specimen, which antedates most well known *Australopithecus,* is a much more advanced form of man. If this is the case, then the well known specimens of *Australopithecus* must represent relicts of an earlier line. These were then existing alongside of more advanced hominids. Some evidence of *Australopithecus* dates at about 4.5 MY, but these early specimens are only scraps. This sort of find, the new Leakey specimen, will undoubtedly be repeated over the coming years and will certainly alter some of the current concepts on the course of man's evolution.

In figure 52 the principal groups of men represented by appropriate taxonomic names, their geological ranges, and ages in years are shown. In this section, we will look briefly at the kinds of fossil evidence upon which this classification is based. In the next, these data will be used as a basis for consideration of the evolution of primates.

Primates originated early during the history of mammals from insectivorous ancestors. The earliest known fossils include a few specimens from the Cretaceous and many from the Paleocene. The group thus dates back at least 70 MY. The early specimens consist of teeth and jaws. They have all come from North America, but this may well be due to accidents of preservation and discovery rather than a restricted range. It is possible that North America was the original center of evolution, but there is no really strong evidence for such a conclusion.

The very early primates were small, somewhat lemurlike, insectivorous animals. During the Paleocene and Eocene, a rapid radiation took place, leading to a variety of lemurlike animals and to others related to the living tarsiers. Records are good in the sense that many specimens are known, but poor in that but a few well preserved skulls and skeletons have been found. One reason for this is that early primates were forest dwellers, as are their few close relatives that still exist today.

The next stage of primate evolution is recorded in North Africa, in the Fayum Desert of Egypt. Most finds consist of teeth, jaws, and parts of skulls of primitive monkeylike animals. Among these is *Aegyptopithecus* which appears to be on the line that led both to apes, the pongids, and to man and his close relatives, the hominids. It

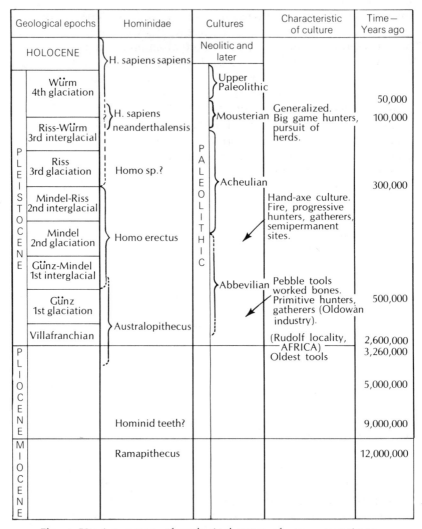

Figure 52. A summary of geological stages of very recent times, the ranges of the various hominids, the cultures associated with fossil men, and the general nature of these cultures. The dates in years are approximations, and the dotted bracket extensions show probable but uncertain ranges.

seems probable that the stock of the Old World monkeys and that of hominoids, pongids, and hominids, were separated by this time. The age in years was around 23–25 million, and the geological age was Oligocene.

Following the Egyptian primates in time, as far as hominoid evolution is concerned, is the genus *Dryopithecus* which ranged from early Miocene to early Pliocene in age. This genus includes specimens which show considerable variation and pass from primitive to more advanced over the time span. Various generic names have been given to different representatives, but the tendency currently is to consider these all as species of one genus. The important point is that the Miocene and early Pliocene epochs have yielded a series of closely related apelike animals which fall fairly close to the dichotomy of pongids and hominids.

Dryopithecus itself appears to be an ancestral ape, already somewhat divergent from the hominid line. It was widespread, being known from Africa, Europe, and Asia. Some living apes probably arose from it. Other fossil apes of this time, *Oreopithecus* of the upper Miocene and *Gigantopithecus* from the lower Pliocene, resemble hominids in many respects. Occasionally, each has been placed within this family, but the resemblances appear to stem from parallelism rather than from direct relationships.

The oldest generally accepted hominid is *Ramapithecus* (fig. 47). Somewhat similar specimens described under various names, including *Kenyapithecus* and *Bramapithecus*, have come from India and Africa. The oldest record goes back about 14 MY. The fossil evidence that hominids were in existence at this time seems reasonably firm. Only teeth and jaws of *Ramapithecus* are known, and it is on these that hominid assignment has been based. Biochemical evidence based on an assumption of constant rates of change of albumins places the separation of man and apes at only about 4–5 MY ago. Reconciliation of the two sets of data, fossil and biochemical, probably will be made, but as yet this has not been accomplished. Here we will assume the fossil evidence that hominids arose 14 or 15 MY ago to be correct.

All pre-*Homo* hominids, except *Ramapithecus,* may be grouped as australopithecines, and the current tendency is to place them all in a single genus *Australopithecus*. Except for very fragmentary remains some 9 MY old, all known specimens of this genus are less than 5 MY in age. There is, thus, a major gap in the record. As a result of exhaustive searches, hundreds of specimens of *Australopithecus* are known. Although most are teeth, a number of skulls, jaws, dentitions, and partial postcranial skeletons have been recovered. The first specimen was found in 1924, so that this accumulation has been made in the relatively short span of 50 years. Most of the fossils have come from the interval between 3 and 1 MY ago, that is, from the late Pliocene and the early Pleistocene. Earlier specimens are extremely fragmentary.

Australopithecines are known only from Africa and on that continent from South and East Africa. Probably only a small part of their full range has been sampled. The absence of such hominids from beds of Pliocene and early Pleistocene age in other continents, even where such deposits have yielded well-developed mammalian faunas, has strengthened the concept that Africa was the center of evolution after the *Ramapithecus* stage. It seems quite certain that there were no New World hominids at this time. Whether or not new finds, possibly tapping as yet untouched environments, will reveal them in the Old World beyond Africa, in particular in southern Asia, can only be a matter of speculation.

Homo erectus is a genus and species now considered to include hominids more or less intermediate between *Australopithecus* and *Homo sapiens*. A number of previously named fossil men, of which *Pithecanthropus* and *Sinanthropus* are the most familiar, are now included in *Homo erectus*. If the find by Leakey noted earlier turns out to be a man of the level of development that preliminary study suggests, it too may belong to this species. *H. erectus* individuals were taller and more perfectly bipedal than *Australopithecus*. They possessed larger brains and relatively smaller faces and dentitions. There were many variations on the central pattern; some specimens are rather close to *Australopithecus* and others closer to *Homo sapiens*. The varieties, to some degree, appear to be distinctive of particular geographic areas, suggesting some degree of subspeciation.

Fossils have come from Asia, Africa, and Europe. Much of the best-known finds of *H. erectus* have been in Indonesia and in China, near Peking. This type of hominid was widespread over the continents of the Old World, probably from about 1.5 and possibly 2.0 MY ago to about ½ MY ago. If the new specimen from Lake Rudolph turns out to belong to the species, the range will be extended to 2.7 MY ago. Both the geographic range and age overlap those of *Australopithecus*. No traces of *H. erectus* have been found in the New World.

Three main sites are known in Indonesia (Java). They range from perhaps 2 MY to ½ MY in age. They have yielded assemblages which include skulls, jaws, dentitions, and postcranial bones. Early specimens appear to be about the same geological age as *A. habilis* at Olduvai in Tanzania and do not differ markedly from it in structure. Later ones indicate considerable evolutionary change, including increase in brain size.

The finest remains have come from China, from Choukoutien, where a series of deposits, perhaps ranging from about eight- to five-hundred-thousand years in age, produced about a dozen skulls, mandibles, many teeth, and some postcranial material. On the whole, the specimens are similar to those from Indonesia. These materials were lost in the course of shipment intended to preserve them during the Chinese war. It is still hoped that they may eventually turn up.

African remains have come from East Africa, at Olduvai in Tanzania, from the Lake Rudolph area, from North Africa in both Algeria and Morocco, and possibly from South Africa where the presence of *H. erectus* is somewhat uncertain. At Olduvai, from bed II, which is 1–¾ MY in age, and from the more recent bed IV, have come two specimens which are probably referable to *H. erectus*. Mandibles with well-preserved teeth, associated with a *hand axe* industry (see fig. 52) much like that at Olduvai, have been unearthed in Algeria. Similar but less well preserved fossils have come from Morocco.

The situation in Europe leaves much to be desired. In beds of middle Pleistocene age in southern Europe, several sites with tools, animal remains, and remnants of fireplaces are known, but they lack hominid remains. These indicate the typical hand axe culture and are about three-hundred-thousand years old. Fragments of hominids have come from Germany and Czechoslovakia. These lie at the base of a frustrating "gap" extending from about three- to one-hundred-thousand years ago. This is a time that has produced tantalizing finds which are scattered, incomplete, and difficult to decipher. Much of the problem seems to arise from the fact that Europe, where the most intensive work has been done, lay marginal to the main geographic habitats of man during the ice ages.

About 75-thousand years ago, the classical neanderthal man, *Homo sapiens neanderthalensis,* was present in Europe. This was during the early part of the last glaciation. He persisted until about 45-thousand years ago. Excellent records of this subspecies of *Homo* have been intensely studied. The populations tapped in western Europe portray but one type of man, probably only one of several types that were existing at that time. Europe probably was sparsely populated and provides a somewhat biased look at the fossil history of *Homo* as a whole.

The European record prior to about 75-thousand years ago was, with only a few finds of materials, deposited between three-hundred-thousand and 75-thousand years ago. Extensive controversy has surrounded the finds that have been made. Samples are small and widely scattered, and preservation is poor; thus, definitive interpretations are difficult.

As studies of man progressed from western Europe to the east and south, fossils that resemble modern man more closely than does neanderthal proper were encountered. From several sites in Africa have come well-preserved but sparse remains which seem to show affinities to types of *H. erectus* represented in earlier beds of the same area. Probably the oldest specimens definitely assignable to *H. sapiens sapiens,* or at least individuals not distinguishable on the basis of what is preserved, come from Mt. Carmel in Israel. They overlap the European neanderthal in time and are associated with similar cultures. The remains from this area, which cover a time span of several thousands of years, show a great deal of variation, some being somewhat neanderthaloid and others more like *H. sapiens sapiens.*

Work in Omo, Ethiopia has brought forth specimens which are modern-appearing in most respects, although archaic in some particulars. The age of the deposits is uncertain but appears to be around 80- to 100-thousand years, very old in view of the *modern* appearance of the specimens.

About 35-thousand years ago, modern man, *H. sapiens sapiens,* appeared in many areas. Although specimens resemble some of those from earlier times, the associated culture is much more advanced. Perhaps 20-thousand years ago, although the time is much disputed, man penetrated the New World and spread through it rapidly. From the time of introduction of modern man and his advanced culture, evolution has been primarily cultural rather than physical. Different cultures have developed in different areas, and studies have come to concentrate upon their natures and interrelationships and not upon the physical nature of the populations through which they have evolved and been transmitted.

Interpretations of hominid evolution

The general course of change

In this section, we will summarize some of the current ideas about the course of human evolution. Much is necessarily speculative and probably will change over the coming years. There are many areas in which the conscientious students of man's history have strongly disagreed and have conflicting opinions; some of these differences are outlined in the following pages.

The monkeys and apes are often termed *simians*. Their ancestors, presimian primates, presumably were monkeylike arboreal animals. In adaptation to this environment, they became active and agile with excellent muscle coordination and stereoscopic color vision. Within the line leading to simians and eventually to hominids, the brain became relatively large and complex, presumably in coordination with

INTERPRETATIONS OF HOMINID EVOLUTION

development of special arboreal traits and in conjunction with increasingly complex social behavior advantageous for group activities.

Presimians and early simians fed mostly in arboreal habitats, consuming fruits, leaves, and other soft vegetation. Partial shift to ground living was undertaken by various primate lines, leading to such partial quadrupeds as gorillas among the great apes and to highly effective quadrupeds such as baboons. Hominids especially emphasized life on the ground by utilizing many features gained in arboreal life in the exploitation of the new environment.

A critical step in the evolution of hominids was this shift to ground life, and it presumably was related to utilization of a new source of food. This involved harsher foodstuffs, vegetation, and perhaps small animals as well. Dentitions in the very early hominids, or possibly prehominids, known from *Ramapithecus* reflect this change. The way for bipedalism had been set up by the *arm-swinging* locomotion of hominids in the arboreal habitats. The animals practicing this way of life are called *brachiators*. This habit in trees and bipedalism on the ground freed the hands for manipulation of objects, both in feeding and many other activities. The possibility of use and fabrication of tools came into existence.

The shift in feeding environment carried the stock across an evolutionary threshold comparable in most respects to some of those we have seen in other instances (see pages 83–97). Thereafter, also in the usual way, the new environment was exploited within the limitation posed by the fundamental biological organization of members of the family. The important point is that, in this regard, hominid evolution is entirely comparable to that of other radiating groups of organisms.

From initial ground feeding, presumably while they were still partially arboreal forest-living creatures, the hominids moved into open woodland and eventually to savanna grasslands. The course from forest to grasslands was common to many mammalian lines at about this time, that is, during the Miocene. One of the best known transitions is that of the family Equidae, the horses, as the stock passed from a forest-browsing way of life through a series of steps to completely plains-living animals. This is shown in comparison with the hominid phylogeny in figure 51.

It was the emergence of grasslands and prairies during the Miocene that made such transitions possible. The various herbivores—horses, antelope, and some camels—were followed into the grasslands by carnivores, particularly the dogs, or canids. Among these carnivores, however, were also the hominids which, as they developed a distinctive pattern of hunting, became increasingly effective predators upon the grassland game.

Modern man was the eventual product of this initial shift, which had paved the way to development of a bipedal terrestrial hunter and gatherer. Once bipedalism had been established, as it seems to have been early among the hominids, most subsequent structural changes were relatively slight. Body size increased notably, three- to four-fold in volume or weight. Bipedalism was perfected by relatively moderate changes of the skeleton and jaws, and dentitions were modified in accomodation to new foodstuffs. The basic physical pattern, however, had been set early, and the lack of pronounced changes thereafter has resulted in a very coherent familial group.

The major change in hominid evolution—increase in brain size accompanied by reorganization of brain components—was closely related to the behavioral aspects of hominid evolution, precisely to those things which set man distinctly apart from all other organisms, even from the most closely related of the great apes. Many, probably most, of the behavioral characteristics of man were consequent upon his becoming a hunter and gatherer of food in open country. A constant interplay of the selective pressures occasioned by these new behavioral patterns and the widening horizons of behavioral potentials as selective modifications appeared is the keynote of the pathways of hominid evolution. This intricate network, involving many feedback systems, is far from fully understood as yet. As new fossil evidence comes to light, and as man's behavior and its genetic basis become better known, changes in viewpoints and interpretations will certainly occur. The general outline of events presented above gives a coherent interpretation which follows from what is currently known.

Very early events

At the stage that pongids and hominids separated, probably in the early Miocene, there was little difference between the members of the two lines. One continued primarily in the forest and the other moved to the ground, to woodlands, and finally to the open lands. Divergence was thus initiated by ecological separation and was accentuated as time passed.

The record of this stage is scant, as we have already seen. The only traces of early hominids are the few specimens referred to *Ramapithecus*, and even these are considered to be prehominid by some students. The areas in which the fossils have been found and the associated mammalian faunas suggest that these hominids were forest-living creatures. The dentition indicates that moderately coarse food was eaten, food probably obtained on the ground. There is strong wear on the first molar, less on the second, and little on the third. This indicates a considerable time lapse during the eruption of successive molars. From this interpretation comes the probability that individuals had a rather slow development and a long period of maturation. Parental care probably persisted over a fairly long period of time.

The absence of any bones of the postcranial skeleton means there are no direct clues on locomotion. There has been much speculation, based mostly on what the structures of later hominids tell about their ancestors. One interpretation is that *Ramapithecus* was an arboreal animal which used its arms for suspension in trees while it fed mostly on the ground. On the other hand, using the gorilla as a model, it has been suggested that it was a *knuckle-walker*. All such interpretations are highly speculative at the present time.

From about 11 MY ago, the date of the most recent specimen of *Ramapithecus*, the record is almost a blank until about 5 MY ago, or for a period of some 6 MY. During that time, a shift to the very manlike bipedal ground-living animal, *Australopithecus*, had taken place. The gap is big and must be filled before we can speak with confidence of the events of this transition.

Later events

Australopithecus There seems to have been two lines of development during at least the last half of the known range of *Australopithecus*, for which the record is fairly good. One included creatures that formed the lightly built *Australopithecus africanus–Australopithecus habilis* line, and the other included heavier animals, the *Australopithecus robustsus–Australopithecus boisei* lines. It is thought that the members of the latter were more adapted to woodlands than were those of the former. It is presumably from the *A. africanus — A. habilis* line that *Homo* arose. It must be realized that with the current information, much of this theory is speculative. There has been and continues to be a great deal of disagreement upon how many distinct types of hominids actually existed and how they were related to one another. What is given here is but one of several possible interpretations.

Australopithecus occurs in association with tools, and it is logical to suppose that he made and used these tools. This has been questioned on the basis that his brain may have been insufficiently developed for such accomplishments. The tools, in this event, are supposed to have been used by some unknown more advanced contemporary hominids. The idea has been given little credence, for it is not, the argument goes, brain size but brain organization that is most critical in performance of peculiarly human activities. Although there is no positive evidence, it is logical to think that the organizational features of the brain of *Homo* had been initiated by the time of *Australopithecus*.

The oldest known tools go back about 2.6 or 2.7 MY. These have come from the vicinity of Lake Rudolph, East Africa. It is in this area that the finds of what appear to be a rather large-brained hominid of the same age have been made. The tools are rather like those of somewhat more recent times found at Olduvai in association with *Australopithecus*. These are part of the *Oldowan culture* and collectively have been called *pebble tools* (fig. 52). These include a fair variety of implements used, it would appear, for pounding, breaking bones, for scraping skins, and probably for hunting. Flakes of worked stones have been found along with the tools at Olduvai.

The making and use of tools and the use of language are two outstanding hominid characteristics. It is reasonable to suppose that the two, which probably were developed coordinately, lay at the base of development of the complex behavioral network which is distinctive of *Homo sapiens*. It can be argued that these skills were initiated in a social structure based upon cooperative hunting and were necessary for successful hunting and transmission of the accompanying culture to successive populations. Cooperation in small hunting groups, division of labor between sexes, and establishment of temporary habitation sites almost certainly were outgrowths of the hunting culture. An essential aspect of cultures in which learning is critical to perpetuation is persistence of motivation, of conscious prediction of the long range results of such activities as the fabrication and utilization of tools. Parental care and teaching become of primary importance. Development of language as a key to cooperation and transmission of information probably was closely related to the initiation and persistence of many of the aspects of the early hunting culture developed by *Australopithecus*.

Much of this is speculative. *Australopithecus* does appear to have been a toolmaker. He hunted animals which formed a part of his food, and he established temporary living sites. It is also known that if the *A. africanus–A. habilis* line is a valid sequence, the brain size increased from about 400 cc to 600 cc without increase in body size. Increase in brain volume almost certainly involved differential development of various parts of the brain, with emphasis on areas related to speech and manipulative abilities. These skills in particular would be of strong selective value in development of a culture such as then envisaged, figuring initially in the physical aspects of the culture and in its preservation. The changes outlined, some well authenticated and

INTERPRETATIONS OF HOMINID EVOLUTION

some speculative, seem to be those that brought the evolving hominids to the threshold of the genus *Homo*. *A. habilis* can be as readily referred to *Homo* as to *Australopithecus* and this, in fact, was its initial assignment.

Homo erectus   In physical characteristics the various specimens assigned to *H. erectus* lie between *Australopithecus* and *H. sapiens*. It is important to recall that *H. erectus* includes a varied array of specimens which have been obtained from widely separated areas. The brain, which was larger than that of *A. habilis* with a mean volume of about 950 cc, ranged from about 800 cc to about 1300 cc, overlapping that found in some members of *H. sapiens*. The total body size was greater than that in *Australopithecus,* being as much as double in volume in some instances. Bipedalism was essentially perfected, and the postcranium was very like that of *H. sapiens*. The skull, however, was rather long and low, that is, *platycephalic,* and the face was sloping. Teeth were relatively large, and canines were somewhat interlocked when the jaws were closed. The head and postcranium evolved at somewhat different rates, displaying a common feature of evolution which produces creatures which are a mosaic of the *old* and the *new*.

H. erectus was a species of advanced hunters. As seen in figure 52, they had a culture designated as Acheulian and sometimes called the *hand axe* culture. The tools were relatively crude, although advanced over those of the *pebble tool* culture. Hand axes, fairly advanced tools, have been found in western areas—Europe, Africa, and western Asia— but not in eastern Asia. Fire and semipermanent habitation sites were characteristic at this stage. The culture clearly was advanced over that of *Australopithecus*. It is not clear, however, just what the nature of the break was. At Olduvai, in the upper level, Acheulian tools are present. It has been argued that these represented tools of an advanced hominid which moved into the area. Also it has been argued that they represent techniques developed within evolving populations which lived continually at the general site. A great deal remains to be learned.

The extensive variation of *H. erectus* over its range suggests the existence of geographic races or subspecies. Hominids are very mobile, but also local populations remain sufficiently isolated that they developed distinctive characteristics. The record for man is insufficient for interpretation of patterns of migration and the extent of gene flow right up to the threshold of written history. The differences between geographic groups of *H. erectus* probably are comparable to those between races of modern man. Part of the problem of evaluating the meaning of such differences in *H. erectus* comes from the fact that remains of the species range over a long period of time. Varieties, perhaps subspecies, are thus temporal as well as geographic. Correlations between the different areas are not sufficiently good that the two factors can be clearly separated.

From these distributions has emerged a much argued point, that of the source of the different groups of *H. sapiens*. Did the different geographic races develop independently from *H. erectus* in several areas, or did *H. sapiens* originate in a single area and spread widely, displacing more primitive men in the course of this spread? Controversy is in part spurred on by possible racial implications, and objective judgements tend to become clouded with subjective emotions.

Probably neither explanation is completely correct, and each may be partially right. Where there is temporal continuity, as in Asia and Africa, the later hominids do seem to resemble *H. erectus* of their own area, indicating possible continuous development. The Australian aborigines, for example, show many features suggesting direct origin from *H. erectus* of the Malayan region. *H. sapiens neanderthalensis* of western Europe probably developed from ancestral forms of the region. It may well be that this was characteristic, but it is also clear that migration of populations was a factor. As matters now stand, empirical data are insufficient to resolve the problem of what the precise course of events may have been.

Neanderthal man   The term *neanderthal* is frequently used to refer to the *classical* neanderthal man of western Europe. This is *H. sapiens neanderthalensis,* characterized by massive superorbital ridges, a sloping forehead, and a flat cranial vault (platycephalic). The mandible is heavy and lacks a chin. Postcranial elements are rather heavy and have various peculiarities not found in *H. sapiens sapiens*. This is a distinctive and coherent group. Its members have often been pictured as primitive, brutish creatures, stooped and almost subhuman. This is certainly a mistake; they are associated with the Mousterian stone tool industry (fig. 52) which includes sophisticated tools, they were big game hunters, and they buried their dead. The subspecies existed during the first part of the most recent glaciation in Europe where it successfully faced severe climatic conditions. The range was from about 75- to 45-thousand years ago.

The classical neanderthal has assumed an unwarranted central position in formulation of concepts of the evolution of man for the reason that most early studies were European. Studies focused upon this one group which turns out to be quite distinct from most other contemporary populations of *Homo.* The remaining members of the genus prior to the origin of *modern man,* approximately 35-thousand years ago, have also been called neanderthal, but this appellation tends to give a false idea of many of their features. Some, it is true, do have characteristics of the classical neanderthal in incipient stages. Others show almost no signs of this trend of development.

Between the time of *H. erectus* and the origin of strictly modern man, the records are very spotty and open to many interpretations. Specimens range back at least two-hundred-thousand years and are known from Europe, Africa, and Asia. At the upper end of their range, they overlap the classical neanderthals. In Java, specimens called *Solo man* resemble *H. erectus* in many ways, but have brain capacities well above the mean of about 950 cc, up to 1300 cc. Solo men probably arose from *H. erectus,* and the line between the two is somewhat arbitrary. Some students have argued that they are in fact *H. erectus,* and it is on this basis that reports of large brain capacities of this species have been based.

Homo sapiens sapiens As we have seen, this subspecies may go back as far as one-hundred-thousand years in time. *H. sapiens sapiens* is characterized by a short high vaulted skull, a flat face, small brow ridges, and small teeth, including noninterlocking canines. To the present time, this species marks the end of physical evolution by *Homo.* There have been no changes in fundamental characters over the last 35-thousand years. Brains are no larger than in neanderthals, either the classic or the less distinctive forms. What has accounted for the particular cranial and dental features of *H. sapiens sapiens* is not well understood. It has been suggested that as more effective tools were developed, the canines and incisors were less used for working such things as hides and became reduced. Coordinately, strong biting and grinding action, producing more efficient mastication of food, developed.

It does seem to be the case that morphologically modern man appeared well before the advanced cultures did. About 30- to 40-thousand years ago in Europe, a culture, termed *Paleolithic* (fig. 52), appeared. It was associated with men who hunted herds of large game, and it has been suggested that new social patterns developed as this way of life was initiated and perfected. The source of this new culture is not known, and the stimulation for many of its characteristics is uncertain. It has been envisaged as accompanying the final physical evolution of man, yet men physically indistinguishable from those with this culture had been in existence for many tens of thousands of years prior to its appearance. It has also been suggested that life under harsh glacial conditions was the stimulus for development of complex social systems and practices, these being necessary for survival. Thereafter, the argument continues, a spread took place into areas occupied by less stressed populations which were overrun.

Men of the modern type did spread over many parts of the world, whether from one center or several, and the less culturally advanced, as well as physically more primitive, men were largely displaced. From that time on, the changes have been basically social and cultural and have progressed at an accelerating pace. Different cultures have emerged in different areas, always with some interchange between partially isolated populations. Gene flow was sufficiently restricted that racial characteristics were preserved, perhaps accentuated, but sufficiently ubiquitous that the subspecies *H. sapiens sapiens* persisted throughout the world.

What happened to the older types is uncertain. They may still exist in the Australian aborigines as relatively little changed descendants. Probably many were eliminated in contacts with culturally superior men, those with better tools and more effective social organizations. Probably hybridization diluted populations so that their distinguishing characteristics were no longer evident. Very clearly, there have been migrations and repeated contacts among different kinds of men. As new areas were settled, new characteristics emerged in adaptation to the circumstances encountered. But these were minor and did not change the fundamental features of *H. sapiens sapiens*. At most, racial differences emerged with the races remaining consistently interfertile.

Unanswered, as noted earlier, is the question of the degree of isolation of the different phyletic lines, that is, the question of whether *H. sapiens sapiens* developed concurrently in several areas or in but one area. In the first instance, it is possible that the races emerged separately, each with a long independent history. In the second, races developed in quite recent times, as modern man spread from a center into areas to which adaptive adjustments were made. These two points of view are illustrated in figure 53 which sums up the two ways of looking at the last stages in man's evolution to the present.

Figure 53 (*opposite page*) Two concepts of the origin of races of *Homo sapiens sapiens*, modern man. (See the discussion in the text.) *A* shows origin from within the species *H. sapiens sapiens*, with the development of an *advanced* man about 50,000 years ago and rapid migration, with adaptation to local conditions in various parts of the world. *B* shows independent development of the races in different parts of the world from stocks of more archaic men placed in the species *Homo erectus*.

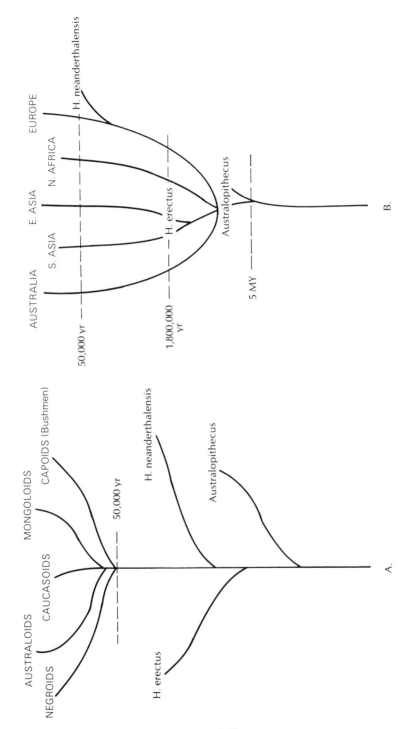

Although modern man has changed little physically during the last 35-thousand years, his impact upon the world has been immense. One aspect of this is his effect upon the evolution of other groups of organisms. Few have been untouched by man's activities, many lines have become extinct either by his direct or indirect intervention. Man, being a newcomer on the scene, developed characteristics to which a great many kinds of organisms have not been able to adapt. Others, rats and mice, for example, have thrived in association with man. Still other lines have been directly affected by man's activities in domestication of animals and selection and hybridization of plants for agriculture. We tend to view all of these activities as *unnatural*. Yet they are little different in kind than some of the activities found among social insects. However they are viewed, man's activities in altering the course of evolution would appear to be immensely greater than those of any other species or group of species since the beginning of life on earth.

IMPORTANT CONCEPTS

AEGYPTOPITHECUS: upper Oligocene from the Fayum Basin of Egypt. Skull and jaw fragments, teeth. Common ancestor of the several species of *Dryopithecus*. Family Pongidea.

AUSTRALOPITHECUS: lower Pleistocene of Africa. Parts of skull, jaws, teeth. On the line of evolution leading to *Homo* and placed in family Hominidae. Contains four valid species—*A. africanus, A. habilis, A. robustus,* and *A. boisei. Australopithecus* has also been called *Plesianthropus* and *Zinjanthropus*.

DRYOPITHECUS: lower Miocene in Africa; middle Miocene to middle Pliocene in Eurasia. Jaws, teeth, one skull. Includes species ancestral to modern pongids and to modern hominids. Also called *Proconsul*. Family Pongidae.

GIGANTOPITHECUS: middle Pleistocene of India and China. Mandibles and teeth. Relationships to other forms unclear. A very large ape—family Pongidae.

HOMO ERECTUS: middle Pleistocene of Indonesia, Africa, Europe, and China. Skulls, mandibles, teeth, postcranial remains. The widespread relatively large-brained precursor of *Homo sapiens*. Also called *Pithecanthropus* (Java Man), *Sinanthropus* (Peking Man), and *Heildelberg Man*.

HOMO SAPIENS (?): earliest remains attributable to the species, occurring in the middle Pleistocene of Europe concurrently with *Homo erectus*. Fragmentary skulls, jaws, many teeth.

HOMO SAPIENS NEANDERTHALENSIS: upper Pleistocene (Riss-Würm interglacial) in Europe, Africa, and Asia. The precursors of *Homo sapiens sapiens* associated with the Mousterian tool tradition. Also called *Solo Man, Phodesian Man,* and many others.

HOMO SAPIENS SAPIENS: *Modern Man*. First seen in the late Upper Pleistocene of Europe and continuing to present times on all continents.

RAMAPITHECUS: Late Miocene to early Pliocene in Eurasia, Africa, and India. Jaws and teeth only. Derived from Dryopithecines, ancestral to the family Hominidae, usually placed in family Hominidae. *Bramapithecus* and *Kenyapithecus* often included in this genus.

REFERENCES

Brace, C.L.; Nelson, H.; and Korn, N. 1971. *Atlas of fossil man.* New York: Holt, Reinhart & Winston, Inc. p. 150.

Dolhinow, P., and Sarich, V.M., eds. 1971. *Background for man*. Boston: Little, Brown & Co. p. 415.

Howell, F.C. 1965. *Early man*. New York: Time-Life Books, Life Nature Library. p. 200.

Oakley, K.P. 1961. *Man the Tool-Maker*. 5th ed. London: British Museum of Natural History Publication.

Pilbeam, D. 1972. *The ascent of man: an introduction to human evolution*. New York: The MacMillan Co. p. 207.

Simons, E.L. 1972. *Primate evolution: an introduction to man's place in nature*. New York: The MacMillan Co. p. 322.

Young, L.B., ed. 1970. *Evolution of man*. New York: Oxford University Press, p. 648.

10 Cultural evolution—man

Why hasn't man evolved?

Homo sapiens sapiens, or man as used in this chapter, has existed for at least 50-thousand and probably more than 100-thousand years with no fundamental physical change. Although many small modifications have arisen, especially in partially isolated populations at the racial level, evolution has not been sufficient to produce a new species or subspecies.

As one views man's history in perspective, 100-thousand years seems like an awfully long time; and it is reasonable to wonder at man's lack of evolutionary change in this interval. Perhaps, indeed, biological evolution does not apply to man at all. However, the fact is that persistence of species with little or no change for such a long interval is normal. One-hundred-thousand years is considerably less than one percent of the 14 or so million years of evolution of the family Hominidae. Man's lack of significant change over his known history is not out of step with the general rates of change in the great majority of other animals for which something is known of the duration of species.

The stability of man's physical structure appears unusual, because during this period rapidly accelerating change in the structure of his society and culture was taking place. Physical stability contrasted with such changes seems anomalous, for in no other evolutionary sequence, either of plants or animals, has anything like cultural evolution occurred. At first glance, it may appear that physical and cultural phenomena should change together, but in fact there is little or no connection between the two. The closest approach to a similar condition is found among some of the social insects, but even in these the nonphysical changes are of lesser magnitude and are accompanied by some physical changes, especially in the development of castes. Even though it is largely nongenetic within species, some genetically related caste evolution has occurred in successions of insect societies.

The initiation of particularly rapid cultural change in man, about 35-thousand years ago, was discussed in the preceding chapter. Forms of written language arose only some five-thousand years ago, providing a base for written history and further accelerating change. Successively to the present time, new ideas and inventions have originated and penetrated, albeit irregularly, through human populations. These have provided many new environmental circumstances to which we might expect physical Darwinian adaptation. Extrapolations of man's rather poorly understood evolutionary history superimposed upon changes in technological environment have been grist for science fiction writers for years. The standard product is a *human* with a huge head (brain) and vestigial legs and feet. Darwinian selection of course is still operative in man and is expressed in minor adjustments to some of his long-term innovations. For the most part, however, this kind of selection has been overshadowed by rapid and pervasive evolutionary changes of other sorts which can be lumped together as cultural evolution.

Cultural evolution has produced a kaleidoscopic series of changes to which man might have adapted through Darwinian selection and to which populations actually have responded to some extent. The need to adapt biologically, however, has been strongly damped by the fact that man can shape his own environment to fit his needs; the necessity for physical change is minimized. Furthermore, man has attained the ability to direct selective processes, supplanting natural selection, and has been extremely conservative in this regard in contrast to the intense selection he has utilized in breeding domestic animals. Added to this is the fact that, by medical means, maintenance of genotypes which would not have survived under natural conditions has thwarted a fundamental process by which natural selection operates. In view of these factors, even the normally slow rate of biological change to be expected in any species, including man, may fail to manifest itself, so that the species may persist indefinitely with little or no change.

THE NATURE OF CULTURAL EVOLUTION 221

Cultural evolution has been the dominant mode of evolutionary change in man during the known history of the subspecies. This situation had led to the idea that cultural change has fully replaced Darwinian evolution and that the future is to be reckoned in these terms. For the time being, at least, biological evolution has taken a back seat to cultural evolution and man's ability to direct the course of his biological change by conscious selection. It is thus important to know something of what culture is and how it evolves. Radical change, a global disaster, for example, if sufficient to immensely reduce the human population, could effectively wipe his current culture from the face of the earth. However, unlike evolution dependent upon genetic continuity, cultural evolution can cease or change radically without major changes in or complete eradication of the bearer.

The nature of cultural evolution

What has evolved?

From the beginnings of *modern* man in the Upper Paleolithic, the general course of cultural evolution can be followed with some confidence. The period to the present has been marked by ebb and flow of complex changes, made intricate and difficult to interpret in detail by man's global mobility. Cultures which developed independently intersected and interacted, and their products introgressed into areas of older cultures to interact with structures produced by still other syntheses. Yet culture as a whole does show a generally constant directional change, so that the term *evolution* can be aptly applied. The interaction of three partially distinct aspects of the totality of culture may be sorted out for anlaysis: (1) social structure, (2) tools and implements, and (3) concepts and ideas. The three are integrated within a framework of what may be termed *human knowledge*. Each has changed notably during man's history, and each, of course, has strongly affected the other two.

Social structure, as we saw in chapter 8, is characteristic of many nonhuman populations. To a limited extent, implements in the form of tools that extend or augment motor activities also are used by animals other than man, especially by the primates. Man, of course, has made immensely greater use of his tools, and they have increased his physical capacities many times over. One way of classifying the cultures of man is based on the production and use of types of tools. The types of tools and their functions have changed, or evolved, forming the basis of technology and, in man's very recent history, initiating the industrial revolution. Man also has developed instruments that extend his sensory capacities—telescopes, microscopes, radio, X ray, and the electron microscope, among others. Now he has successfully undertaken extension of his capacities to do mental work by the development of electronic computers.

Social structures and implements have truly evolved also, but at the heart of cultural evolution lie the changes in the realm of concepts and ideas. From the beginning of written history, and probably long before, man has endeavored to explain himself and the world in which he lives. Explanation has taken many forms, ranging through crude mythologies to polytheistic and monotheistic religions and rationalistic and scientific forms of explanation, more or less in this order of temporal succession. Self-consciousness, an ability for abstract thought, and the capacity to transmit abstract concepts to others are necessary ingredients for the evolution of ideas. Biological evolution endowed man with these capacities, but their realization, which forms the foundation of cultural evolution, proceeded independently of biological evolution.

How culture has evolved

It is easy to draw appealing analogies between the courses of biological and cultural evolution. This holds especially true for the changes in social structure and implements. Implements are an outcome of applied knowledge, and the development of this kind of knowledge also exhibits many parallels with biological evolution. Phylogenies of the *branching tree* type can be constructed, with *specialization* characteristic of the items at the ends of the branches (see figures 54 and 55 for examples). Occasionally, as in biological evolution, some highly specialized feature is *preadapted* to a new use and forms the basis for a new development in another *environment*. Radiation may follow, and perhaps extinction. Divergent change, parallelism, and convergence are common phenomena in cultural history. A *key innovation* such as the optical lens provided the base of an extensive radiation in its utilization; gunpowder lies at the beginning of a great radiation of explosive compounds with multitudes of uses.

Social evolution has occurred in somewhat similar ways. During the waning of the ice ages, some 10–15-thousand years ago, man spread rapidly to many parts of the earth not previously inhabited as warming took place to open new habitable areas and open pathways between areas. With this spread, many partially isolated populations

THE NATURE OF CULTURAL EVOLUTION

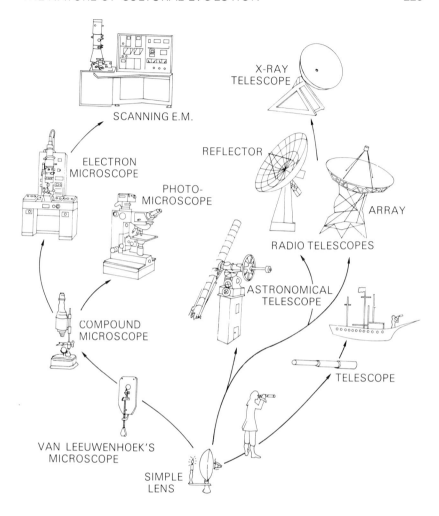

Figure 54 A phylogeny of mechanical aids to men, starting with the simple lens and branching in two directions, one bringing large distant objects closer and the other making very small objects visible. Many of the *factors* of evolution can be seen—radiation, adaptation, selection, extinction, and so on. However, there was a controlling intelligence in the evolution pictured here; man was in back of the successive steps, if not of the planning of the full course.

developed, following their own ways of life in particular environments. Cultural adaptations to particular ways of life—fishing and

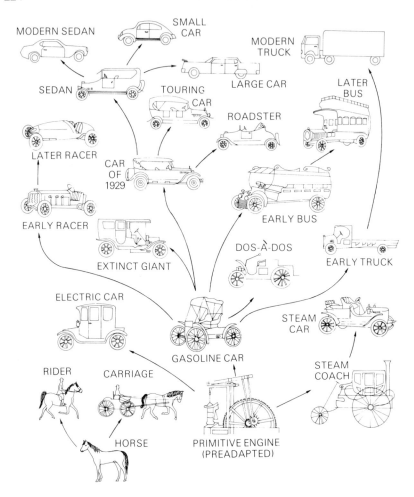

Figure 55 The evolution of automotive transportation. Note the several adaptive types—hauling, mass transportation, individual transportation, sport. Note the extinctions of the "dinosaurs" and the primitive mode of transportation by horses which has persisted in special or primitive environments.

hunting in the north, nomadic life in the deserts, or agriculture in more fertile regions—resulted in considerable divergence of social structures. Upon coming to inhabit separate but similar environments, populations with a common social heritage underwent cultural changes which were similar, or parallel. Even with the absence of a common heritage, strikingly similar cultures developed in isolation, converging upon one another. In America, from Mexico to Peru, and in southern Asia, from Mesopotamia to China, development of agriculture and animal husbandry was accomplished with a resultant concentration of peoples and the evolution of remarkably similar civilizations.

The evolution of concepts and ideas, except for some of the offshoots of applied knowledge related to social structure and implements, does not fit this pattern as neatly. Discussion of it will be deferred for the moment while we turn to the fundamental differences between cultural and biological evolution. These rest primarily in the way that information is transmitted from one generation to the next. In biological evolution, of course, transmission is genetic. In cultural evolution, transmission is accomplished strictly by learning. Learning by imitation and, it appears, by a primitive form of *teaching* takes place in many nonhuman animal groups, producing at least a rudimentary social continuity. Such learning also is important in man, but the full fruition of culture throughout history has depended upon its perpetuation by teaching of a new generation by its progenitors and by peers within the established generation. The use of tools, mythologies, mores, and abstract ideas of many sorts are all transmitted by teaching. During the earlier stages of man's cultural evolution, teaching was largely by example and word of mouth; later, writing became important and provided a means for greatly expanded, more pervasive, and more permanent continuity of cultures. Regardless of the means of transmission of abstract ideas, oral or written, some form of articulate, symbolic language is necessary. It is only in man that this has developed to any great extent.

Its methods of transmission give cultural evolution some unique properties. It has a *Lamarckian* aspect, that is, features acquired by one generation can be passed on directly to the next. This occurs in biological evolution only when the new acquisition is a change in the genetic materials. Furthermore, cultural evolution is readily reversible; it can and often does revert to an earlier state. Reversal in biological evolution above the level of the simplest gene mutation is so highly improbable that it can be disregarded. Extinction of a biological sequence, some phyletic line for example, means the elimination of the genetic basis of the line and, as a result, the death of all the phenotypic bearers of this information. Extinction of a culture may also occur, but some of the individuals who carried the culture may still be in existence after its extinction.

These differences, which are real and have sweeping and pervasive consequences, make it dangerous to carry the analogy between biological and cultural evolution beyond superficial levels. Yet, as we will see in the following chapter, this has frequently been done, especially in the areas of morals and ethics. These philosophies represent particularly sensitive and delicate aspects of man's culture, falling more in the realm of ideas than the realm of social structure and implements where analogies hold greater validity.

The evolution of concepts, ideas, and knowledge

This topic deals with a controversial area marked by widely divergent opinions. It touches upon the field of the theory of knowledge, *epistemology*, that lies at the heart of many philosophical disputes. Concepts and ideas do evolve, however, and in some gross sense, knowledge has accumulated. Knowledge is a broad term which involves both facts and inferences. Applied knowledge, one part of the total, draws from both of these sources and, as noted earlier, is basic to the development and perpetuation of the physical structure of society. Through implementation it feeds back to areas of inferred knowledge, forging a strong link between knowledge in general and other areas of human culture. It, too, evolves.

Facts, in a broad sense, are interpretive and depend upon the prevailing concepts and ideas of a society for their validity. Such an elemental *fact* that the sun rises in the east is *fact* only if we have a coordinate system and a particular concept of the sun-earth relationship. It is, of course, specifically false under our current understanding of the relationships and motions of the two bodies. An idealist may maintain that the page you are reading or the tree that you see does not, in fact, exist as such but is merely a construct of your mind. That physical objects do not exist as we know them *out there* has been a point of considerable argument. The realistic or common-sense approach argues that they do exist, but this argument requires a particular conceptual base for its acceptance. Facts, then, are relative and, like truth, exist only in the conceptual framework, or the *paradigm*, of the observer. It is the evolution of the conceptual aspect of knowledge that is most important and upon which we will concentrate.

Whereas evolution of social structure and implements has tended toward specialization and particularization, concepts and ideas have tended to evolve to become more inclusive and general. They arise from isolated sources and become progressively better integrated. As K. Popper has pointed out in his essays on *objective knowledge*, knowledge has come from many stems which, over time, have evolved toward convergence into a single trunk (fig. 56). Stated

THE NATURE OF CULTURAL EVOLUTION

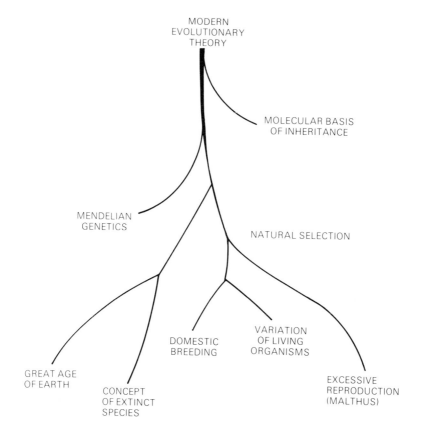

Figure 56 An example of a *phylogeny* of knowledge, actually a reversed phylogeny. The particular leads to the general. The theory of evolution and some of the main contributing ideas, as they merge, are used for illustration.

otherwise, man in his continuing quest for explanations has tended to grasp for increasingly encompassing concepts that integrate ever larger suites of the *facts* of the world in which he lives.

Two dichotomous evolutionary pathways have developed during man's history, each dependent upon a different assumption concerning the proper way to attain knowledge. In both branches, there exists a common source of information in the form of some very general set of *a priori* hypotheses. One of the two branches leads to an elaboration of knowledge whose source is conceived as divinely inspired and which forms the basis of mythologies, religions, or other dogmas in which the ways of interpreting the world are dictated by *revelation*.

Idealism is a slightly different form of this kind of knowledge. The main characteristic of *inspired knowledge* is that its concepts are not subject to empirical test; their appropriateness is judged by the criteria of internal consistency and how well they fit the dogma.

From the evolutionary point of view, the most important aspect of this branch of the dichotomy is that the encompassing concepts may arise rather rapidly in the form of some dominant central theme, as in various mythologies, and persist over long periods of time, eventually becoming formalized into laws. Man, it would seem, is a creature who needs mythology, who needs goals and heroes, and who needs something to worship; his need is satisfied by such an inspired creative conceptual framework. While truth is fundamental and unalterable within any single system, new systems with new truths have arisen many times in man's history. New sets of truths may replace the old. Polytheistic faiths often have been replaced by monotheistic systems, and schisms have been basic to sweeping revolutions. Progressive change in this area often has been saltatory as a result of sudden nonevolutionary modifications; this feature distinguishes its progress from that of Darwinian evolution.

The second branch of the dichotomy includes acquisition of knowledge under concepts of materialistic empirical science. Broadly speaking, this is the field of *realism*. The most important aspect of concepts in this branch, the theories and hypotheses, is that they are subject to empirical testing. The inferences which derive from theories and hypotheses can be tested directly against observations of the *real* world, the world of common sense. *Facts* are conceived under the tenets of realist philosophy. This approach to explanations of man and the universe was formalized quite late in man's history, although in an informal way it lay at the base of man's commonsense approach to everyday living from the beginning. Observations of nature produced not only explicable events but also many that were not immediately explicable; the latter provided a basis for nonmaterial explanations characteristic of the other branch of the dichotomy. At this level the two forms of knowledge approach a merger.

The evolutionary pathway of scientific knowledge, like that of idealistic knowledge, has led in the general direction of broadening integration of concepts by synthesis of data from many different sources. It is the nature of scientific knowledge to be but an approximation of the *truth*, however defined, and that its explanations tend to be only momentarily satisfactory. This condition exists because of selection which, in this area of cultural evolution, results mostly from tests and criticism. The selected, that is, successful, theories are those that have withstood continued testing of their inferences against observations. A second selective mechanism has been the degree of inclusiveness of a theory; the broader or more extensively explanatory a theory is, the more it meets the goal of explanation. Similarly, simplicity is a primary criterion in selective evaluation of theories.

An important aspect of the evolution of concepts in the scientific motif is that explanations of one problem create new problems not merely unrecognized ones, which did not exist prior to solution of the first problem. In this way, problem solving by theory-inference-objective testing, given man's urge to explain, has *built-in* evolutionary mechanism. New problems and ideas emerge from the structure of the solution applied to the initial problem.

K. Popper has illustrated this by a simple example. Once enumeration in the form of 1, 2, 3, 4, 5, 6, 7, 8, 9,... was formulated, the problem of odd and even numbers and their relationships came into existence, and a wealth of problems emerged. Among them, for example, is the question of the infinity of integers and the problem of infinity in general. If there is an infinity of numbers, is there not also an infinity of both odd and even numbers; that is, are there not classes of infinity? How often do we find the solution of a problem followed by the statement "but that creates a new problem?" This is a critical and independent aspect of the evolution of knowledge in the scientific mode.

At the heart of cultural evolution, irrespective of the branch we follow, lies the creative role of concepts and theories. The major changes that have come with time have emerged with man's continuing desire to explain, and his broadening of the concepts involved with that explanation. Of course, conflicting ways of gaining knowledge have existed and still exist; they form the base of most philosophical conflicts among differing schools of thought. The conflicts serve the useful purpose of stimulating critical analyses in opposing areas. Such a basic conflict is at the center of the disagreements between religious and scientific concepts. In their pure forms as sources of information, these branches of the dichotomy of the evolution of knowledge are incompatible; by softening or ignoring some of the basic tenets of each, some degree of compromise can be reached, but this can result in only partially satisfactory rationalization.

The remarkable impact of Darwin's explanation of evolution was due directly to the casting of his mechanistic explanation of the nature of organic change, his testable theory of natural selection, into a world which viewed life in all its ramifications in a long-accepted theological content based on revealed truth. During the century in which Darwinian doctrines have been important in man's explanation of himself and the world, many alterations of our concept of the universe have taken place. Some of these were initially shocking, but they have now become an essentially subconscious framework of the way in which we view ourselves and our relationships to the biological and physical world in which we live. In the final chapter, we will look into some of the most important aspects of these changes.

IMPORTANT CONCEPTS

ABSTRACTION: the formation of a concept or idea apart from particular events or material objects. An idea formed from a combination of words or other symbols which cannot be directly expressed in the physical world.

CIVILIZATION: the totality of culture of a people, distinguished by a particular government, art, technology, and way of life and thought.

CULTURAL SELECTION: the differential survival of ideas, art forms, tools, forms of government, etc. The selective agents involved are numerous and complex, but may include tests and criticisms, simplicity, usefulness, conventionality, pragmatism, attitudes, or inclusiveness.

CULTURE: the totality of the physical and mental products of a society, including its ethics, art, tools, concepts, religion, government, institutions, beliefs, laws, literature, etc.

EPISTEMOLOGY: that branch of philosophy that deals with investigations of the origin, structure, methodology, validity, and expression of knowledge.

EVOLUTION, LAMARCKIAN: the idea, proposed by Lamarck, that changes acquired during the lifetime of an organism could be hereditarily passed on to its offspring. Although this idea is invalid for biological evolution, it can be tenuously extended to cultural evolution in which knowledge and skills acquired during an individual's lifetime can be passed to the offspring through teaching.

EVOLUTION, SALTATORY: the idea that evolution, in this case cultural evolution, can proceed by quantum jumps, rather than by gradual, directional change.

KNOWLEDGE, APPLIED: the application of what is known to practical ends; manifested as industry, research, technology, etc.

KNOWLEDGE, INFERRED: information obtained by reasoning, and based upon real objects and phenomena but extending beyond them. The ground upon which the intellect lands after an inductive leap.

KNOWLEDGE, INSPIRED: knowledge from an intuitive, extrasensory, or divine source that is not based on events, objects, or processes in the *real* world.

LANGUAGE, SYMBOLIC: a means of expression of communication that may be written down in the form of signs or characters. It applies particularly to the symbols which express visually, and in permanent form, a spoken language.

RACE: any of the major biological divisions of man, usually applying to Mongoloid, Negroid, and Caucasoid groups; but imprecise common usage has charged the term with emotion and left it virtually meaningless in a technical context.

SOCIETY, HUMAN: a group of human beings living together in the same area, and forming a homogeneous unit that exhibits stability through time. Usually distinguished by a particular culture.

TEACHING: the means of assuring continuity of culture from one generation to the next by transmission of written, spoken, or demonstrated information. To make a *very* broad analogy, it is the *hereditary* process of cultural evolution.

TECHNOLOGY: the science dealing with the development and improvement of the physical implements used by a culture. Can also refer to the totality of the implements themselves.

REFERENCES

Hawkes, J. 1965. *Prehistory*. New York: New American Library. p. 476.

Popper, K. 1972. *Objective knowledge: evolutionary approach*. Oxford: Clarendon Press.

Wooley, Sir Leonard. 1965. *The beginnings of civilizations*. New York: New American Library Inc. p. 636.

11 The impacts of evolutionary concepts

Perspectives

Evolution in a most general sense embodies the idea of descent with change. We have studied this extensively as it applies to biology and more briefly with respect to culture and have seen that even ideas may be said to *evolve*. Here, we will be concerned with the evolution of evolutionary concepts themselves and with the ways in which they have come to occupy a critical part of our scientific and philosophical understanding of the universe and man's place in it. Trends of thought during the last two centuries have been accompanied by the gradual incorporation of the idea of evolution into the patterns of everyday life, so that now we habitually think in terms of gradual progressive change. One need only browse through a sample of current newspapers and magazines to recognize the ubiquity of our sense of history and its unfolding; we base our understanding of immediate events upon causes found in the recent past and predict the future on the basis of these and events of the present.

It is difficult to conceive that people at other times could think differently, but of course they did, and it is not improbable that still other modes of thought will be embraced by man in the future. Even now, evolutionary thought is centered mainly in the Western world, although with the spread of western culture it has been carried widely abroad where its impacts have been felt intensely in some areas but hardly at all in others. Recognizing this, we will, for simplicity, concentrate on events in the United States and Europe in the rest of this chapter.

Darwin's publication of *On the origin of species* in 1859 was one of the critical milestones in the development of current ways of thought and of social patterns and mores. However, Darwin's book was one of a series of crucial events that were to radically alter human thinking; for this reason, it is useful to examine briefly the ideas that preceded *The origin*, since without them the book would not have been possible. Like *The origin*, each of the important steps in the evolution of our ways of thinking was a product of its time and accomplished a synthesis of the experience and ideas of the years just passed, and was, in some ways, revolutionary. The gradual accumulation and evolution of ideas forms a fascinating story that becomes bound up with the personalities of those who were responsible for them; and it is therefore unfortunate that our account must be such a brief one.

As a supplement to our treatment of this material, page references to the *Columbia history of the world* (Garratt and Gay 1972) are cited at the end of the chapter. In addition, there are numerous references to the *Darwin* volume (Appleman 1970) for special topics relevant to the events preceding and immediately following publication of *The origin*. More general ideas are covered in two excellent books by George Gaylord Simpson, *The meaning of evolution* and *This view of life;* the first is written from a scientific viewpoint, the second from a philosophical one. Taken together, these books provide an insight into the point of view of one of the outstanding evolutionists of our time.

The stage for *The origin*

Copernicus was on his deathbed in 1543 when his *De revolutionibus orbium coelestium* was published. He could have had little understanding that this single work would have such pervasive influence that the beginning of the scientific revolution may be dated from its appearance. The book itself was a dry technical report of his research in the area of astronomy. It combined mathematics, which had an ancient heritage, with observations to document a new way of viewing the mechanics of the solar system. The resulting synthesis expressed a fundamental point of view that was to be incorporated into most later scientific works in all fields.

Copernicus' conclusions, removing the earth from the center of the solar system and the universe, caused no more than a few ripples until they were elaborated upon by others. Galileo in paticular accepted the concept of a moving earth and used it as a base for his own astronomical investigations. It was not until the work of Kepler and Newton, however, that the Ptolemaic concept of a central earth surrounded by celestial spheres was discredited and the hold of Aristotelian physics at last was broken. Static concepts of the earth and cosmos, in which the idea of evolution was unthinkable, were replaced by a new understanding of motion and the laws governing it.

Newton, in his *Principia mathematica philosophie naturalis* (1786) combined observation and analysis (empiricism and rationalism) in a sophisticated fasion to set the course for all science that was to follow. Along with Galileo, Descartes, and Kepler, he presented the physical cosmos as a machine powered, according to Newton, by the force of gravity.

The concept of a dynamic universe replaced the Aristotelian philosophy that had become a part of Christian doctrine under the influence of St. Thomas Aquinas during the thirteenth century. The older doctrine was naturalistic and, in this sense, fitted what was to come much later. It prevailed until the seventeenth century as a critical part of church doctrine, carrying with it the concepts of permanence and stability. The Copernican interpretation of a moving earth came into direct conflict with these basic Christian tenets, and later, under the duress of the Inquisition, Galileo was forced to concede its heresy, even though the concept was basic to all of his own work. The conflicts between religion and science begun during this period are yet to be resolved, although they have passed through phases ranging from near compromise to open hostility.

Stemming from the scientific discoveries and the concepts they generated were the social and theological modifications of the seventeenth and eighteenth centuries. The new and radical ideas of science and the concurrent changing perspectives of man and society during this *Age of Reason* laid the foundations for development of ideas about individual rights, equality of all men, and the kind of liberalism that was eventually to lead to democracy. Voltaire was the main spokesman of liberalism at the time, but the roots of his thought are to be found in Newton. Denis Diderot's great encyclopedia, a monument to liberalism, science, and the good of the people, carried the revolutionary doctrines to a wide segment of the literate population.

The critical, rational, and logical temper of these times, while providing a base for social reform, also strongly affected theology; one of the results was a natural theology called *Deism*. It was a *scientific* theology and, as such, was highly analytical. Locke and Voltaire both were strong adherents to deistic faith. God existed under the tenets of Deism, and atheism was strictly forbidden; but He became a supreme mechanic in the mechanistic universe, a sort of Newton with the power to create. No room was left for theistic morals, but morality was rationally necessary and had to be found by man searching within himself for the moral sense created there by God.

Deistic theology-philosophy was based on the conviction, from science, that man's knowledge could come only from his senses; according to this materialistic or positivistic viewpoint, understanding was to be attained only by analysis of information. The duality of mind and body envisaged by Descartes was abandoned in favor of a mind-body unity that had a machinelike quality. The Newtonian *cosmic* machine was matched, by analogy, with the *mind-body* machine.

To one side of Deism was the still strong orthodox religion which was powerful particularly because of its role in education. On the other side, emerging from the rational approach to knowledge, was strict materialism and atheism. These three threads ran concurrently through the intellectual life of the eighteenth century, repeatedly intersecting and coming into confict. The last mentioned reached its logical epitome in the *skepticism* of David Hume, who distrusted both theology and science. As an agnostic, in the sense the term is used today (one who does not know), he found that the lack of ability to demonstrate the ultimate truth of anything required a faith in man, in the individual, and in his role in the world of nature. For Hume, religion was an important factor in human experience, but it was a man-made affair—naturalistic, not supernatural.

Reaction to cold analytical Deism was not long in coming; it was crystallized in the thinking of two persons in particular. Immanuel Kant in his *Critique of pure reason* took off from Hume and Jean Jacques Rousseau, who preceded both Hume and Kant in time. Kant saw the *reason* of the scientist as a practical rather than a pure sort of reasoning. Thus, by denying its fundamental precept, he pointed the way out of the dilemma of skepticism. Both religion and science found new breath in his analysis.

Rousseau was of great importance to the movement which culminated in the so-called *romantic age*. He swung sharply away from his early affiliations with the deists to take a *practical* view of man; he tried to discover what faith or religion man, as man, would find most satisfactory. Rationalism was not the whole answer, for it left no place for the sense of being a man and an integral part of nature. Morality, he felt, must have its base in feeling and emotion, not in some causal system. The base of Rousseau's concept of man lay in nature, and could not be approached by reason alone; he could not see nature as simply a path to knowledge of the Creator.

At this juncture, a number of factors which were to lead to evolutionary thinking had come into being. People were beginning to base their thinking upon observation, analysis, and a naturalistic worldview. However, old, cherished ideas are a long time dying, and concepts of unity of the whole, stability and permanence, still held the stage.

Under the influence of Kant, Rousseau, and their predecessors who were ill-at-ease with the Deistic analysis, the Age of Reason gave way to the romantic age. This period began at about the time of the French Revolution in 1789 and lasted until 1848 when extensive revolutions on the European continent consummated a growing disillusionment with the free-swinging approach to life, literature, the arts, and theology. The romantic age, while it lasted, was a time of great creativity and expression of genius in various forms. It was the time of Scott, Wordsworth, Keats, Shelley, Byron, Goethe, and Pushkin in literature, of Turner and Delacroix in art, and of Beethoven, Chopin, Schubert, and many others in music. Scientists continued to work unperturbed, and made great strides in physics, chemistry, and astronomy. As the philosophical retreat from a cold mechanical world took place, however, the importance of the world of organic beings began to emerge. At about the same time, there was an outbreak of piety and evangelistic fervor; the massive clashes of theology and science of earlier times, however, were minimized because life and abstractions were treated as separate entities. Conflicting analyses, very simply, were no longer in vogue, and many earlier problems were no longer considered important.

Reason was not abandoned during this time but was put into the context of experience, part and parcel of the total experience of man and nature. In such an emotional and intellectual environment with its strong overtones of natural philosophy, the concept of evolution could develop from the parallel courses of art, philosophy, and science. Evolution as a *form of thought* was born during the romantic age, and during the first part of the nineteenth century it became the most common form.

As early as the eighteenth century, men such as Maupertius, Diderot, and Comte de Buffon were evolutionists and held many of the ideas later synthesized by Darwin into a coherent theory. It was

Maupertius who first had the germ of the concept of particulate inheritance, later to be called Mendelian inheritance. Lamarck, likewise, was an evolutionist and proposed a method of change from generation to generation by inheritance of characters acquired during the organisms' lifetime. Erasmus Darwin, grandfather of Charles, suggested that the will and needs of organisms could be a source of heritable variation. These men were the main forerunners of Darwin, but their ways of thinking were slow in gaining acceptance.

By 1800, geologists had come to recognize the great antiquity of the earth and paleontologists and amateur fossil-hunters had discovered the remains of vast numbers of animals and plants in ancient sediments. It had become evident that the geological processes currently active on earth were adequate to explain events of its history, and that the earth displayed an orderly evolutionary development—not long periods of stability marred by periodic catastrophes. In 1833 Charles Lyell published his *Principles of geology,* a book based on thoroughly evolutionary philosophy as it applied to the earth. Educated persons had no problems with evolutionary ways of thinking, but the Deity continued to play a definite role in many minds, even though faith had been periodically shaken by intellectual events since the Middle Ages.

During the same time, physical science was busy learning to harness the energy sources of the earth that would soon permit rapid industrial expansion in Europe and North America. A confidence in inevitable progress was emerging. The turmoil of genius, freedom, social unrest, revolution, and piety as well, ran its course and began to wane after 1848. The ideological philosophies, which stemmed from Kant, persisted but often were much modified. They included a strong historical consciousness best expressed in the works of Hegel. The idea of evolution thus remained strong, even though much of the cold materialism of science was rejected. Later, during the twentieth century, analysis was again to become dominant, but this would come long after the intense romantic feeling of the early nineteenth century was well in the past.

The past 150 years have seen a continuing breakdown of orthodox Christian concepts as formulated by St. Thomas Aquinas in medieval times. However, until the middle and late nineteenth century, there had been a strong tendency to adjust new ideas and concepts to fit within the framework of orthodoxy. As might be expected, this was particularly true with respect to man, his origins, and his place in the universe. The ideas of transformation of species (evolutionists before Darwin were called *transformationists*), while often suggested, were not widely accepted. That the universe depended ultimately on a deity, though denied by the atheists, was still the dominant theme among the great majority of people, including the scientists.

The mystical quality of life, the finished *perfection* of organisms and their precise adjustment to their roles in natural communities lent an air of incredibility to the idea of organic evolution. Species could become altered and varied, as breeding experiments demonstrated, but the idea of whole new kinds of organisms originating gradually from preexisting kinds seemed incredible. Could such perfect structures as the vertebrate eye, the human brain, or the closely coordinated elements of a bird's locomotory system have come into being without intelligent design by a creator? The outwardly compelling evidence for design and purpose formed a strong buffer against transferring the evolutionary nature of the physical world to the biological and social world, particularly as it applied to the origin of man.

Biological change, during the first half of the nineteenth century, was generally regarded either as change toward *perfection*, as in ontogeny, or as degeneration, as in ill-formed variants of a species. The fall of man and his expulsion from Eden was a degenerate change. However, with Buffon, Lamarck, Erasmus Darwin, Lyell, Chambers in his *Vestiges of creation*, Charles Darwin in his research prior to 1859, and Spencer and Wallace all providing evidence for gradual progressive change in the organic world, the conceptual dam was ready to break.

Effects of *The origin*

It was into this climate of confident economic progress, where struggle determined success or failure of competing enterprises and individuals, and where there was wide acceptance of an historico-evolutionary way of thought despite preponderantly Biblical conceptions of man and his origins, that *On the origin of species* was published by its understandably reluctant author, Charles Darwin. The first printing was sold out on the day of publication.

Biological thought, of course, was immediately affected. Resistance to the transformation of species and to natural selection as a mechanism of change was initially strong in some quarters, but soon the concepts were generally accepted, at least by the scientific community. Other effects of *The origin*, those with which we will be concerned in this chapter, were in part overt and immediate and in part subtle and enduring. Among the former was reintensification of the conflict between evolution and theology and the phenomenon that would come to be known as *Social Darwinism*. Related to theology was the age-old problem of ethics which had now to be cast in the context of evolution; even today there is no stable resolution.

Among the more subtle impacts of Darwinism were its influences on philosophy and the general bases of everyday thought as reflected in human behavior. These have permeated current societies so thoroughly and have become so integrated into our thought patterns that they are barely recognizable as offspring of evolutionary theory.

We will concentrate on theology, ethics, and philosophy as being most important in present-day society. In addition, we will examine Social Darwinism, which rose and fell but has left deep marks upon our society. All of these areas are closely related and there is much overlap; however, they are treated separately here to simplify analysis. And although emphasis on these few areas leaves out other important ones, the basic ingredients of the impacts of evolutionary theory on modern life, biology aside, are to be found in just those three areas.

Theology

The conflict between science and theology relative to man's origin and his position in the universe rose to a fevered pitch with publication of *The origin*. The antagonism, as we have seen, was of long and bitter standing and had taken on a number of forms. Biology, despite pre-Darwinian support for transformation of species, had played a relatively small role in the conflict. Theology, on the other hand, had been a strong factor in retarding the development of evolutionary thought. Even Spencer's presentation of his evolutionary philosophy (Appleman 1970, p. 489) had reserved a place for a somewhat remote God in his relegation of matters such as final cause to an *Unknowable*. Darwin's mechanistic explanation of the origin of species hit home sharply.

Darwin, less in his accepting of evolution than in his supplying a means of change independent of design and purpose, flew directly in the face of orthodox theological doctrines of fixity, final cause, and perfection. Now evolution, too, was given a machinelike character. Yet Darwin, for all of this, was a deeply religious man. His challenge to orthodoxy worried him, as he implied in *The origin*. It was not until 1871, in his *Descent of man*, that he applied evolutionary concepts to man. Nevertheless, oblique statements on the subject in *The origin* were immediately noticed by orthodox theologians who saw man's origin as set forth in Genesis seriously imperilled.

The reaction from some quarters was pointed and vociferous: "Darwin endeavors to dethrone God;" ". . . if Darwin is right . . . then the Bible's teaching in regard to man is utterly annihilated;" "These infamous doctrines have for their only support the most abject passion;" or, in reconciliation, "Supernatural design produces natural selection."[1] The arguments ebbed and flowed between Darwin's opponents and supporters with few holds barred. Many scientists, notably T.H. Huxley and the famed geologist Charles Lyell, welcomed Darwin's ideas. Others, among them the distinguished anatomist Sir Richard Owen and the prominent zoologist Louis Agassiz, dissented vigorously. The paleontologists in particular tended to look upon evolution with disfavor. Darwin himself had been troubled by the geological record and he had devoted considerable space to it in *The origin*. Today this very imperfect record of past life is still used on occasion as an argument for creation.

Over the years, an uneasy compromise began to emerge as evolution in one form or another became permissible within the doctrines of many religions. Unavoidably, the very personal immediate nature of God gave ground before concepts of a more remote sort of deity. Various solutions, some with roots that go back well before Darwin, have been proposed. One interprets the accounts in Genesis as allegorical, written to suit the tastes of the peoples at the time of their original telling. The essential meaning of God is preserved. The *days* of creation become the ages, and population of the earth becomes gradual with creation taking on an evolutionary flow. A second approach totally separates science and theology, with each treating valid but unrelated areas of man's being; the physical and spiritual duality of man is its basic theme. In a third interpretation, God becomes an impersonal but pervasive force. To some He represents or is the organizing principle back of the order in the universe; others see Him as the order and laws themselves. Final cause is not considered part of the order, and it is pushed to a remote conceptual corner. A variation of this idea is that God was the *prime mover* who set the universe on its course and then withdrew, allowing subsequent events to follow an evolutionary course. Still another approach weds science and theology into an evolutionary theology, a scientific mysticism which has gained popularity in recent years. All of these philosophies still embrace some universal purpose or goal, independent of immediate causality.

---

[1] These quotations and their original sources are given in Appleman (1970) as a reprint of a passage titled *The final effort of theology* by A. D. White. This is a section of *A history of the warfare of science with theology* (1896), New York.

Teilhard de Chardin (chap. 2 and Appleman 1970, p. 458) has elaborated upon the idea of purpose in an interesting way. According to him, evolution is destined to lead unrelentingly and irresistably toward perfection. Inevitable improvement is reminiscent of a similar tenet in Spencer's philosophy, but it is predicated on very different grounds. Both attempt the difficult merger of materialistic empirical science and ethereal extramaterial forces. Although somewhat similar to the natural theology of the eighteenth century deists, the Teilhardian type of formulation assigns a very different and more remote role, both spatially and temporally, to the extramaterial component.

For all of the compromises arrived at since *The origin,* the conflict still goes on. It is now less between the theologians and scientists than between the scientists and laymen of the church. This surfaces as sporadic efforts to ban or restrict the teaching of evolutionary theory to students in public schools, or at least to give equal time to *creation* theory.

More profound and less subject to public debate is the duality which runs through the thought patterns of a large majority of the populations of the Western world. For a great many persons, the moral and social patterns they were taught as children stem from creationist religious doctrines of the Judeo-Christian faith. Yet evolutionary modes of thought dominate the cultures and endow the pervading world-view with an aura incompatible with religious orthodoxy. The inevitable conflict of materialism and mysticism exists, even though it is largely unconscious; and this produces tensions that tear at the ethical and moral structures of many societies. Man's apparent need for some metaphysical or mythological base for his continued well-being, recognized in every naturalistic philosophy, has not been met by evolution or by natural selection as a whole. At the same time, the findings of science have called into question the older theological doctrines by which man lived and have seriously weakened their hold on his faith.

Ethics

The problem of ethics lies at the heart of the relationship between science and theology, and figures importantly in the conflict between the two as discussed in the previous section. During the last century, the orthodox theological concept of good gave way to a naturalistic concept of good—from "if it meets God's will, it is good" to "if it is natural, it is good." The two obviously are not mutually exclusive and have been variously rationalized.

Once Darwinian evolution was accepted, it weakened man's faith in the scriptures and, inadvertantly, scriptural morality as well. Some of the strongest attacks on Darwinism were based on this fact. However, it is not easy to determine the precise role that evolutionary concepts had in influencing the complex social changes of the nineteenth century, particularly with regard to the nebulous area of ethics. The fact that ethical standards vary widely between societies and fluctuate rapidly and unpredictably over the years in any given society makes the task doubly difficult.

One of the best means of evaluating basic social changes is through examination of the governmental documents of a society, for it is in these that the ethical core and punishment for its violation are formulated. In stable societies the laws change slowly, usually lagging somewhat behind general changes in ethical concepts and behavior. Even when ethical changes are the result of revolution rather than evolution, the basic structure of the ethical code and its *legalization* may emerge with only a few fundamental modifications. Hence, the enduring changes appear to have been slow; the rapid short-term modifications are superficial and usually have only minor effects on the broad scope of change. If we consider the timing of appearance of the doctrine of Darwinian evolution in this light, it is seen to have followed rather than preceded the liberalizing trends that pervaded the Western world in the eighteenth and nineteenth centuries. It was in large measure a result, not a cause, of changes in the realm of social and ethical standards.

The dire consequences that many predicted would follow the onslaught of *Godless* Darwinism never materialized in Europe or North America. Godlessness, or *state atheism*, did arise in the course of the revolutions that resulted in formation of the communist states, especially in the Soviet Union. Although they seemed to bear out the predictions, the communist revolutions actually were only minimally affected by Darwinism. The revolutionary doctrines of Karl Marx, applying largely to the economic sphere, stemmed more from the evolutionary dialectics of Hegel; it was Engels who structured Soviet science to conform to the doctrines of Marxism. Marx, however, sought confirmation of his ideas in the parts of Darwin that were supportive of his credo; these were minor, and he drew very little substance from the main ideas put forth in the first editions of *The origin*.

From these several considerations, it appears that, at least superficially, the effects of Darwinism on ethics may have been relatively insignificant. There are two matters in the relationships of ethics and evolution, however, that do deserve our attention. One is the search for a new base for ethical guidance in evolutionary processes and evolutionary history; the other is related to the consequences of the influence of evolutionary concepts on biology.

With the slow realization of the ubiquity of evolutionary processes in the universe came attempts to derive the basis for an ethical system from evolutionary concepts. Allusions to efforts of this sort crop up time and again in popular and semipopular literature. The evolutionary philosophy of Herbert Spencer, as we will see in the section on Social Darwinism, was centered on a strong ethical theme. More recently, serious scholars have undertaken the reexamination of this problem; among them are Sir Julian Huxley, George G. Simpson, and Ralph Gerard.

Julian Huxley (Appleman 1970 p. 405) offered a dual approach to evolutionary ethics. On the one hand, he derived the basis of all behavior from organic evolution as a whole. On the other, he emphasized that standards of behavior are strictly human and that morality enters evolution only with the evolution of man's consciousness. From this duality he established a series of ethical statements: it is right to realize the possibilities of evolution; it is right to respect human individuality; it is right to construct a mechanism for social evolution that will satisfy the first two conditions. In elaboration of these points, he developed broad fundamental guidelines for human behavior with regard to the future of man.

In *The meaning of evolution* (1949), Simpson dismissed evolution in all its manifestations as an ethical basis for human behavior, pointing out the logical failure of analytical natural ethics. In many ways, he was echoing Thomas Huxley who made similar statements shortly after publication of *The origin*. The problem of naturalistic ethics was tackled head-on by E.G. Moore, a very influential philosopher in the early part of this century; his strongly negative conclusion was that there is absolutely no basis for ethics in naturalism. His opinions stirred up a lively debate which has been going strong ever since.

Simpson found moral responsibility inherent in man's knowledge, not in *nature*. From this idea developed an *ethic of knowledge*. This does not include cut-and-dried formulae for ethical behavior; it merely points out the *need* for ethical behavior and its relationship to evolutionary processes. The ethic of knowledge is derived from human nature and was attained through evolutionary means. Logically, as man and his societies become modified, his values are going to have to change. In this way, the ethic of knowledge is evolutionary. Unlike the Newtonian Deists and their counterparts such as Rousseau, Simpson makes no appeal to a metaphysical ethical base; the responsibility for moral action and judgement rests upon man himself.

In effect, Simpson has said that man, with his burden of consciousness, can and should determine what is best for himself and his envi-

EFFECTS OF *THE ORIGIN*

ronment and carry out the appropriate changes. The modern concern with nature and the environment, while momentarily pragmatic, depends upon just such an ethical sense of responsibility of man to nature and to future generations. How we carry out our responsibility depends in part on evolutionary knowledge, but the directions and aims do not derive from knowledge of either the processes or history of organic evolution.

Both Huxley and Simpson, using evolutionary information, consider the individual to be the most important element in society and repudiate the idea of *superorganism* in which the individual is subordinated to the whole. The concept of superorganism, as we saw in chapter 8 can be successfully applied to colonies of social insects and may have important implications for group or colonial evolution at that level. Applied to man, superorganism has a long history. It has been renewed and expanded in recent times, particularly by the biologist Ralph Gerard. The *state* is usually taken as the unit and individuals contribute to its well-being by their activities; it is their right and their duty to do so. Ethical systems of responsibility to the state are but a short step from this kind of thinking. Under such a system, any individual activity detrimental to the maintenance and well-being of the whole is ethically *bad*. The formulations of Simpson and Julian Huxley and the contrary idea of superorganism reflect the two major conflicting social philosophies of today—individualism and socialism.

A more alarming set of problems has come out of the great increase of biological knowledge during the last decades, culminating in an ability to maniupulate heredity both in man and other organisms. Results are, within reason, predictable. As man's power in this area has developed, a growing sense of need for its responsible use has begun to emerge. The means to control the direction of human evolution are at hand.

So far, much of the application of biological knowledge to humans, largely through the medical professions, has been carried out with little consideration of long-term consequences. Growing awareness, however, has raised an urgent need for careful decisions on a multitude of biologically oriented problems. These include control of population size, abortion and contraception, genetic engineering, cloning of human foetuses, artificial insemination, man's responsibility to all organisms and the physical environment of the planet, preservation or elimination of defective genotypes in human populations, euthanasia, and even the restriction of freedom of scientific inquiry in biology and the physical sciences. These are the outstanding, *sensational* problems, but there are many others.

All such problems have obvious ethical content. Many of them did not exist a century ago, and those that did were much less critical and less in the public eye than they are today. Population control, although it was a recognized issue, did not seem particularly important. Today it is thought by many to be one of the most pressing ethical responsibilities man must face.

Consideration of any one of these problems in depth will expose the current ethical dilemma. This stems, as we have seen, from the weakening of ordained ethics as a guide and the use of *ad hoc* ethical decisions founded on hazy ill-defined concepts of "the good of humanity" and "responsibility of man." Although the trend has been toward man's responsibility and away from dictated ethical choices, most actual attempts to arrive at answers serve only to stimulate renewed conflict between the two alternatives. Except in extreme cases, we operate from a dual standard.

The question of the value of life will illustrate this dual standard nicely. Human life, we mostly agree, must not be taken—"Thou shalt not kill." Wars and capital punishment ride roughshod over this basic ethic, using the rationalization that they, too, have ethical validity; history is full of *holy wars* ("an eye for an eye") and religious and political persecution. Despite the social sanction of certain types of killing, however, there is a strong sense of guilt in most of the people who kill. The sense of moral wrong in killing is so strong in many persons that they will suffer to almost any length to avoid taking a human life. To arrive at an honest objective answer to the question, "Why is it wrong to kill another human being?", is not easy outside the framework of external sanction. An answer is even more difficult to find if the question is extended beyond human life; is it wrong to kill a deer, a mosquito, a bacillus? Most answers are superficially pragmatic, their rightness determined by the circumstances and the outcome, but at some level they must revert to an ethical decision.

The problem of the sanctity of human life relates directly to such current problems as abortion and organ transplantation. In the case of abortion, the right of a woman to do with her body what she will is pitted against the question of whether or not abortion constitutes the taking of a unique life. It becomes necessary to define the precise point at which a foetus becomes a full-fledged human life. If purely human ends, for example, convenience or population control, are put first without consideration of the sanctity of life, the ethical considerations take on a whole new aspect. Ethical and legal difficulties of the same sort arise in establishing criteria for death of a donor prior to surgery for organ transplants. This is a relatively new problem which has arisen directly as a result of great advances in surgical knowledge and techniques.

In all of the aforementioned cases, a cold rationalistic ethic of what is good for humanity is opposed to the supreme right of the individual to do what is best for him. Such ethical chaos is certainly not new; it was resolved in earlier societies by formalized divinely inspired codes of ethics, but this option is not so easy to apply today. As we have seen, concepts of evolution have been responsible for much of the breakdown of traditional ethical codes, but evolution has been unable to supply a moral and ethical basis to replace the one that was lost. In addition, medical sciences and genetics continue to advance over what was once sacred ground. The problems they pose, not the least of which is the direction of the evolution of man in years to come, require ethical decisions if they are to be resolved.

Understanding evolution and its mechanisms has given man the capacity to perceive his condition and to direct his own evolution in such a way that many of the problems he now faces can be reduced or eliminated. Thus, the ethic of responsibility of man, long in existence, is becoming intensified. However, biological knowledge alone cannot draw the ethical guidelines for determining how to exercise this responsibility. The burden of determining long-term ends is an awesome one, and it must rest in man's conscious knowledge of himself and his relationship to this total environment. Once directions are determined, the means to the ends are at hand. Following them, of course, will create new problems which will require new solutions at higher levels as man's self-awareness and his sense of what is right continues to evolve.

Social Darwinism

Evolutionary justification for social structure and behavior followed quickly in the wake of publication of the works of Herbert Spencer and Charles Darwin. New *natural laws* had been found, and it was not illogically assumed that these could make a meaningful contribution to the laws governing society and the individual's relationship to it. The culmination of attempts to meld society and evolution was Social Darwinism, which was in vogue during the latter part of the nineteenth century. Although its glory was short-lived, Social Darwinism has contributed parts of its substance to our own social milieu; these have been so thoroughly assimilated that they are difficult to recognize, but they continue to play an important part in current social dynamics.

There was a brief resurgence of some of the concepts of Social Darwinism during the 1930s in the form of the fascist ideology of Hitler and Mussolini. This was a racist-militaristic interpretation for which the seeds had been sown much earlier, and was but a caricature of true

Social Darwinism. Its development caused worldwide repercussions that are too well known to require elaboration here.

The phenomenon called Social Darwinism is rather badly misnamed. Its roots lie much more in the evolutionary philosophy of Spencer than in Darwin's concept of the origin of species by natural selection. Spencer, who had an engineering background, combined certain laws of physics—in particular the law of conservation of energy—with the results of population studies of Malthus and the idea of progressive evolution. The product was a philosophy of progress in whch the *struggle for existence* and the *survival of the fittest* figured prominently. "Struggle for existence" is a phrase used by Darwin, but in what he called a "large, metaphorical sense," in his exposition of natural selection by preservation of the fittest individuals. Use of the common phrase suggests a strong tie between the doctrines of Spencer and Darwin, but the similarity is more apparent than real. Overt struggle played a relatively minor part in Darwin's interpretations.

Spencer's philosophy was a product of the industrial age and rapid economic growth in England; of course, this was true to some extent for Darwin's theory as well. Progress was characteristic of the time and was seen in Spencer's eyes as the inevitable consequence of the conflict between evolution, which creates diversity, and dissolution, which leads to uniformity.

Developing concurrently with the dialectical concepts of Spencer was the evolutionary doctrine of dialectical materialsim espoused by Marx and Engels. With its Hegelian base and economic emphasis, this doctrine diverged widely from Spencerian philosophy; but the two proceeded from a markedly similar set of assumptions. It is remarkable that they produced two diametrically opposed social philosophies. The capitalistic world of Spencer and the communistic world of Marx, children of a single concept, now confront one another with the threat of atomic warfare and global annihilation.

The United States, expanding industrially and devoted to individualism in a socially conservative society, was fertile ground for Spencer's philosophy. The doctrine spread rapidly and became an active element in the social structure of the nation. The basic idea—applied to individuals, classes, industrial empires, and later to races and nations—was simply that the survival of the fittest, a *law* of nature, provided a measure of worth. This, it may be noted, was compatible with the puritan ethic, so that Social Darwinism could be acceptable to a basically devout society.

The most fit, of course, was the most successful. The only path to success was struggle, and the right to succeed lay in the superiority demonstrated by success. Thus, each individual had a God-given right to his position and carried with it a commensurate responsibility to society. The less able, the poor, the crippled, those of low status, were where they were because they were less fit and had failed in the struggle. This belief, arising with Spencer and his times, still thrives; it is, at most, thinly veiled in the bastions of ultraconservativism in many parts of the world.

Dominance by a few and the idea that dominance was their God-given right was nothing new on the world scene. The novelty of Social Darwinism was that the grounds for justification of superiority were found in natural law, not in divine right or brute force. Because of its dependence on natural law, Social Darwinism is intimately entwined with the ethical concepts and problems discussed in the previous section; indeed, when viewed in perspective, the two may be seen to have *coevolved* during the early part of this century.

The doctrines of Social Darwinism, eagerly grasped by industrialists as rationalizations of both their positions and their responsibilities, have gradually changed, and their roles have become less important. However, the effects of the strong racist-militaristic phase of the decades preceding 1950 are still with us. Since that time, the concepts of evolution have been modified so that natural selection is now very different from what it was in Darwin's and Spencer's day. Survival of the fittest and struggle for existence have little meaning in contemporary evolutionary theory. Fitness is now measured by the contribution of the individual to the gene pool of the next generation. Social Darwinism could not arise or find justification in a climate permeated by modern evolutionary thought. The *tooth and claw* side of nature and its role in change, however, is still appealing and evident enough that the bloodier natural laws tend to remain as *justifications* for many kinds of social acts.

Philosophy

Philosophy to the average person is at the same time very remote and very intimate, depending upon how he is thinking about it. The works of most philosophers are highly intricate and analytical, often couched in a special jargon, and are too tenuous and pedantic for a casual reader; but the broad problems with which philosophers deal are largely refinements of matters that concern all persons of their times. As interests and emphases have changed over the centuries, philosophical problems have become modified in such ways that the old ones have faded away, still unsolved, and new ones have taken their place.

Striking changes in the structure of societies have taken place during the last century, as we have seen. What we would like to do in this final section is to determine the extent of the influence of evolutionary thinking upon the philosophies that moulded and arose from these changes. This is not a simple task because, ideally, we should know all about how people think and what guides their actions. Even if we did, we would be hampered by the fact that many factors have been involved in these changes, and they have not acted the same ways in all societies. If we focus only upon evolutionary influences, ignoring such matters as technological development, the rise of science, increasing ease of communication, or expanding knowledge of the universe, the role of evolution will be much overemphasized. However, we will take this chance and look mainly at evolution anyway, bearing in mind that it was by no means the only influence guiding the progress of philosophy.

During Darwin's time and well before, basically evolutionary philosophies enjoyed widespread popularity. Most influential was the complex and somewhat obscure philosophy of Hegel (1770–1839), including the concept of the evolutionary dialectic of progress through synthesis of opposing doctrines. It is impossible to explore this difficult philosophy of inevitable progress here, but its effects are important to us today in two ways: (1) as the basis for Marxist dialectical philosophy which, since its adoption in the Soviet Union, has persisted as the only sweeping evolutionary philosophy with *built-in* progress; and (2) as the focal point for much of modern philosophy which arose, in large part, from criticism of some of Hegel's points of view.

Other purely evolutionary philosophies, such as those of Spencer and Bergson alluded to earlier, had their day. These were *systems* philosophies through which the cosmos and everything within it is viewed from a single central concept. They have since dropped to insignificance, and today they are essentially dead.

There have been some later systems philosophies, such as that of Whitehead (chap, 2), that were evolutionary but in somewhat different ways. Whitehead's concept of universe as organism is extremely broad but is consistent with the trends of the modern era in being critical. It rejects science's probabilistic conception of the universe as meaningless. The philosophy of Teilhard de Chardin (chap. 2) is, likewise, cosmic in scope. It differs from others in combining science and metaphysics and utilizes physical and biological sciences to comprehend the more elusive motivations of change. This philosophy, despite its mixture of essentially incompatible *hard* science and mysticism, has had rather pervasive influence in recent years.

Most modern philosophers have turned toward a more practical particular approach to philosophical questions. Reliance on experience, on experimentation, on *real* things, and on critical logical analysis is an outgrowth of the methods used so successfully in the physical sciences. Darwinian thought stressed function and process acting within a temporal framework. Taken together, these concepts lie at the heart of modern biology. A suite of critical analytical philosophies are dominant in the Western world today, forging strong links between biology, evolution in its larger sense, and prevalent modes of thought. Although these philosophies are often at semantic odds with one another, they all have common properties and are harmonious with what we know of the natural and physical sciences. To trace them all back to Darwin is to greatly overdraw the issue, but the climate generated by *The origin* was definitely beneficial to their development.

Many of the critical philosophies developed as reactions against the sweeping *Absolute* of Hegel; they preferred to examine particular specific matters and recognized a diversity of entities rather than a cosmic unity or *grand design*. The world could be meaningfully studied not as pre-Darwinians thought, because a static purposeful plan exists to be revealed, but because it has a temporal aspect—a history—which is manifested in natural phenomena and which may be unravelled and reconstructed, because the processes of change have remained about the same for all time. There is no single end or ultimate goal; rather, there are many potential end points that shift with changing events and circumstances. *Relativism* is an important aspect of this changing world whose guidelines are universal physical laws and whose constraints are the events of previous history. The ultimate nature of causation and truth and similar questions are not answered; they simply do not exist as legitimate questions.

As in the area of ethics, *truth* in the realm of pragmatic philosophy lies in the outcome of the process being examined. The correctness of a proposition rests in how well it coincides with observations; essentially, if it works or if it isn't inconsistent with what can be observed and materially tested, it's right. If two different theories yield identical inferences, no decision of relative *rightness* can be made between them. To argue the comparative worth of such theories is trivial and meaningless.

Existentialism, especially in its atheistic mode, centers on the uniqueness of individual experience and focuses pragmatism on the personal level. Man's capacity to make choices on the basis of his experiences is emphasized and, as in many other modern philosophies, developmental aspects (growth of knowledge with experience) are important. Further, it places responsibility for the consequences of his acts directly upon the individual human being, thus making a strong ethical commitment.

Logical positivism, or logical empiricism, is a philosophy that guides the progress of much of modern science. It takes the position that only observation can determine whether or not a statement is factual or *true*. To have meaning, statements must be either after the fact (*a posteriori*), based on experiment and observation, or before the fact (*a priori*), synthetic and predictive. In other words, only empirical and mathematical statements, including statements of symbolic logic, are to be considered valid. Karl Popper's contention that no theory or hypothesis is subject to direct validation but can be tested only by attempts to falsify it falls within the same general pattern of critical evaluation by tests and experiments.

These and other similar philosophical doctrines figure prominently in the intellectual climate of our times. They stem from the fabric of everyday life which provides attitudes, problems, and eventually supplies a feedback that influences and molds the very life that generated them. They are part and parcel of our literature and art as well as our more homely interpretations of minor events of the day. Different animate and inanimate materials and processes, varieties of experimental results, and multitudes of individual observations produce a bewildering complexity of data input; but this can be unified either analytically or by recognition of common underlying principles. Physics and chemistry, once coldly deterministic, have expanded within the concepts of time and function to take on a naturalistic historical cast. Biology, once naturalistic and nonempirical, has taken on much of the discipline and precision of physical science. Society and social processes are recognized as being in a normal state of flux, changing and becoming changed through time.

Underlying the existence of our mode of thought today, we find four predominating themes: (1) change, (2) the interpretability of natural events without recourse to an ultimate design or purpose, (3) the relationship of ends and means in the context of process and function, and (4) the unity of biological and nonbiological processes. These ideas all were presented in *On the origin of species* under the aegis of natural selection. If we sum up these attitudes and see how they appear in our everyday *street* philosophy, we find a set of attitudes which may be termed *modern naturalism*.

What modern naturalism has done is to transfer responsibility from an external source, a Deity, to man himself. Within man alone reside whatever aims and purposes there may be. Man's purpose is based on a recognition that change is natural and inevitable and that its directions can, within limits, be controlled and predicted. As long as we work within the confines of natural law, we feel deep down that we can control most things—even the weather, earthquakes, disease and famine—and that one day we can go to the stars. We certainly have held to the idea of evolutionary perfection in that we feel that change is to be equated with progress, but in recent years the idea of inevitability of progress has ebbed. Evolutionary perfection is an *ad hoc* adaptation involving a particular situation, not progresss to a Teilhardian *omega point*.

Where we go from here is uncertain, for having gained the power to direct evolutionary change and to predict with some certainty the outcome of any course of action we may undertake, we have not been very successful at deciding what we want to do with the future. Until there is some consensus founded in a rational factual base, our knowledge of evolutionary processes and our naturalistic sense of the world cannot fulfill its possible promise. Long before such a consensus can be reached, if history can be taken as a guide, the philosophical bases of our actions may have passed far beyond what now seems to be the mainstream. Evolution, in a very real sense, creates the very problems we are trying so desperately to resolve. Is man, with his growing understanding, equal to the responsibility he has placed upon himself? Only time will tell.

IMPORTANT CONCEPTS

AGNOSTICISM: a theory of knowledge. It may refer to the idea that man is incapable of knowledge of a particular subject or to the notion that man can never attain a knowledge of God.

ATHEISM: very strictly, the idea that God does not exist. More generally, the idea that God exists, but on a remote, nonpersonal basis.

CAPITALISM: an economic system. It is based on the idea that production and distribution of goods should be controlled by privately owned concerns for profit, under competitive conditions.

COMMUNISM: a level of social and economic development. Based on the concept of a classless society in which each individual contributes to society according to his ability, and receives goods and services of society according to his need. Competition, profit, and the nonexistence of the state as a repressive force are characteristic. Communism is at a level beyond that of socialism.

CONSERVATISM: attitudes tending toward preservation of established traditions and toward resistance and opposition to changes in the same.

COSMOLOGY: see Important Concepts, Chapter 2.

DEMOCRACY: a form of government. Government by the majority of the people, either directly or through elected representatives. Acceptance of the principle of equality of rights, opportunities, and treatment for all individuals.

DEISM: an attempt to rationalize religion by establishing reason as the only valid approach to revelation. Both reason and ethics in this scheme are natural phenomena. Knowledge of God is attainable through reason.

EMPIRICISM: see Important Concepts chapter 2.

ETHICS: a discipline dealing with subjective judgements. Tries to determine right vs. wrong, good vs. bad and expresses attitudes of approval or disapproval for particular actions, conditions, or ends.

EVOLUTIONISM: see Important Concepts chapter 1.

EXISTENTIALISM: a theory of knowledge. Based on the idea that there is no dichotomy between the physical world and the psychological world, it states that all natural phenomena have their existence in states of mind. Both the sources and elements of knowledge are sensations in man's consciousness.

FASCISM: a system of government. It is characterized by an inflexible one-party dictatorship which forcibly represses all opposition. Permits private ownership of means of production and distribution of goods but under the watchful eye of the state.

HUMANISM: see Important Concepts chapter 2.

IMPORTANT CONCEPTS

IDEOLOGY: the study of the nature and source of ideas. Can also refer to the doctrines, opinions, and modes of thought of an individual or group.

LIBERALISM: in politics and religion, it is an attitude of tolerance of the views held by others and open-mindedness to ideas that challenge tradition. It emphasizes the personal freedom of the individual.

LOGICAL POSITIVISM: an attitude that emphasizes empiricism and the *scientific method* and seeks a unity of science in which a simple set of interconnected natural laws can apply to all the sciences—physical, biological, and social.

MATERIALISM: see Important Concepts chapter 1.

MECHANISM: see Important Concepts chapter 1.

METAPHYSICS: a branch of philosophy that deals with the first principles, that is, it seeks to explain the nature of being and reality. It is closely related to theories of knowledge (epistemology).

MORALS: the acceptable ethical codes, conduct, and customs of a particular individual or group.

MYSTICISM: a theory of knowledge. The idea that it is possible to achieve knowledge by communion with God and that not all knowledge need come through the medium of human reasoning powers.

MYTHOLOGY: a tradition of stories explaining natural phenomena, the history and origin of man, or certain aspects of culture as seen through the heroic and not-so-heroic actions of men and their deities.

NATURALISM: the idea that the universe requires no supernatural cause or sanction. Man, in this framework, is only an incidental product of natural events in a universe that is self-existing and self-directing.

PHILOSOPHY: see Important Concepts chapter 1.

POSITIVISM: a theory of knowledge. It is based on empiricism and states that the highest form of knowledge is simple description of perceivable phenomena. According to the positivists, man's search for knowledge evolves through three stages—the theological, the metaphysical, and finally the positive.

PRAGMATISM: see Important Concepts, chapter 1.

RATIONALISM: see Important Concepts, chapter 1.

RELATIVISM: see Important Concepts, chapter 2.

SCEPTICISM: a theory of knowledge. It is based on the idea that no knowledge at all, or at least no absolute or unquestionable (perfect) knowledge, is attainable by man.

SOCIAL DARWINISM: a social and economic system. Based on the so-called *evolutionary* principles of *struggle for existence* and *survival of the fittest*, it applies to competition of individuals, corporation, or other entities within the society. The value of an individual or group is judged by its success in the struggle.

SOCIALISM: a social and economic system at a level beyond capitalism and below communism. The ownership of means of production and distribution rests in society or community, not in private concerns. All members contribute to the work of the community and all receive its products according to their contribution and need.

TECHNOLOGY: the realm of application of science to practical or industrial ends. The science of tools, machines, gadgets, implements, and miscellaneous devices.

THEOLOGY: the study of God, religious doctrines, and matters divine. Attempts to understand the relationships of God to the universe and to man.

REFERENCES

Appleman, P., ed. 1970. *Darwin*. New York: W. W. Norton & Co., Inc. pp. 385–422.

Garratt, J.A. and Gay, P., eds, 1972. *Columbia history of the world*. New York: Harper and Row. pp. 380–400; 591–97; 681–92; 707–21; 799–807; 859–71; 1237.

Kuhn, T.S. 1959. The structure of scientific revolutions. *The international encyclopedia of unified science,* 2d. ed. 2: 210.

Menaker, E. and Menaker, W. 1965. *Ego in evolution*. New York: Grove Press. p. 266.

Russell, B. 1959. *Wisdom of the west*. New York: Crescent Books. p. 320.

Simpson, G.G. 1949. *Meaning of evolution*. New Haven: Yale University Press. p. 346.

Simpson, G.G. 1964. *This view of life*. New York: Harcourt Brace Jovanovich, Inc. p. 304.

Index

References to figures are in boldface type. Page numbers for definitions are in italics.

Abbevilian, 202
Abstraction, *231*
Acheulian, 202
 culture, 211
Achillea, 114
Acoelomates, 97
Adaptation
 phylogenetic, 58
 physiologic, 58
Adaptive radiation, 57, 83, *168*
 course of, **14,**
 reptiles to mammals, **90**
Aegyptopithecus, 201, *217*
Aerobes, 52, 63
Agassiz, L., 241
Agnosticism, *254*
Algae, 69, 107
Allele, 121, **122,** 132
Allopatric species, 132
Altruism, 182, *189*
Amino acid, 30, 47, 119
Ammonoids, 86
Anaerobes, 52, 63
Analogy, **137, 140,** *168*
Analysis, 26
Angiosperms, 153–54
Annelida, 97, 102
Ants
 castes, 179
 eusocial systems, 177–80
A postiori, 26
A priori, 26
Aquinas, St. T., 235
Aristotle, 1, **2,** 24, 235
Arthropoda, 97, 103
Atheism, *254*
Atmosphere, primitive earth, 36–38
Australia, 145–46
Australopithecus, 192, 201–04, 209–11, *217*
 skull, **198**
Autotrophs, 69

Bacteria, 120
Bees
 castes, **177**
 eusocial behavior, 175–77
Bergson, H., 1, 21, 22, 250
Bernal, 43

Bernard, C., 18
Bilateralia, **74,** 97
Bioherms, 69
Biology and ethics, 245–47
Binary fission, 57
Biston betularia, 128, 149
Blastoids, **87,** 88
Blue green algae, 54, 57–58
Brachiopods, 97, **102**
 evolution, 84, **85**
Brain
 capacities, 211, 213
 Homo sapiens sapiens, **197**
 chimpanzee, **197**
Bramapithecus, 203
Bryozoa, 97
Buffon, comte de, 237
Bulawayan Formation, 54

Calvin, M., 39
Capitalism, *254*
Carbohydrates, 41, 47
Castes, *189*
 ants, **176**
 bees, **177**
 general features, 175–76
 origins, 180–83
 termites, 180
 wasps, **176**
Catalysis, 41, 47
Cell, 30
Centrioles, 69
Centrosome, 69
Cephalopods, 85, **86**
Chaetognatha, 97
Chambers, 239
Change
 cyclical, 4, 10
 linear, 5, 10
Changing world, 4–5
Character, 132
Chemistry, origin of life, 38–44
Chemosynthesizing bacteria, 55, 69
Chromatids, 71
Chromosomes, 63, 121, 133
 mutations, **124**
Civilization, *231*
Classification, 110

257

Clines, 113 **114,** 133
Cloud, P. E., 81
Coacervates, 42–43, 45
Codon, 119, 133
Coefficient of relationship, 184
Coelenterata, 97, 102
Coevolution, 153, *168*
  plants-larvae, **159**
Colias, 151–52
  eurytheme, 162
  meadi, 106
  philodice, 162
Colony, 189
Communism, *254*
Complexity, 10
Conservatism, *254*
Continental drift, 142
Convergence, *168*
Convergent evolution, 136, **137**
Copernicus, 243–45
Coral snake, 146
Cosmology, 26
Cowania, **161**
Crinoids, **87,** 88
Crossing over, 124, **125,** 133
Cultural evolution
  adaptations, 224–25
  analogies, biological, 222–26
  dichotomy, 227–28
  facts about, 226
  how occurs, 222–26
  inspired knowledge, 228
  scientific knowledge, 228–29
  transmission of culture, 228
Cultural selection, *231*
Cyclical change, 4
Cystoids, **87,** 88

Darwin, C. 8, **14,** 247–49
Darwinian
  evolution, 82, 109, 229
  thought, 251
Darwin's finches, 136, **138**
DDT, 128
Dehydration condensation, 39, 41
Deism, *254*
Deistic philosophy, 236–38, 244
Democracy, *254*
Deuterostomia, **74,** 97
Dialectical materialism, 6, 10, 20
Diderot, D., 20, 235, 236
Dimorphism, 149
Diploid, 63, **64,** 69, *189*
Divergence, 136, 137, *168*
Division of labor, *189*, 210

DNA, 117–21, **119**
  hybridization, 133
  structure, **117**
Dobzhansky, Th., 116
Domestic animals, 130
Donorism, *189*
Dryopithecus, 192, 203, *217*

Echinodermata, **87,** 97, 104
Echinoids, **87,** 88
Echiurida, 98
Ediacaran fauna, 75, **76**
Ehrlich, P. 158, **159**
Einstein, A., 21–22
Élan vital, 21
Electrophoresis, 151
Emergence, 19–23, 26
Empiricism, 26
Endothermy, 129
Engels, F., 6, 19
Environment
  origin of life, 36
Enzyme, 47
Epistomology, *231*
Erithizon, 141–42
Ethics
  Darwinism and, 242–47
Eucaryote, **61,** 69
  algae, 55
  cells, 52, 59–62
Eusocial systems, 173, *189*
Evolution
  a way of thought, 237
  biological, 107
  cellular, 31
  chemical, 31
  complex social systems, 187–88
  convergent, 136–46, **137**
  cultural, 219–30
  divergent, 136–46
  early life, 51–72
  impacts of, 233–56
  Lamarckian, *231*
  mechanistic explanation, 240
  Neodarwinian synthesis, 115
  of concepts, ideas, knowledge, 266–30
  of man, 171–218
  parallel, 136, 146, **137**
  prebiologic, 29, 31
  regressive, 163–65, *169*
  saltatory, *231*
  social structure, 221–22
  tools, implements, 221, **223, 224**
  unit of, 115
  ways of viewing, 5

INDEX

Evolutionary
  ethics, 244
  landmarks, 44
  origin of life, 44–46
Evolutionary theory, 3
  development of, 234–39
  impacts, 233–56
  of universe, 15
Evolutionism, 10
Existentialism, 251–52, *254*

Fascism, *254*
Fig Tree Group, 34, 35, 36
Finches, **139**
Fire, use of, 202
Fisher, R., 129
Fitness, 184, *189*
Flagellum, 60, 69
Ford, E. B., 128, 148
Fossil record, 77-78
Fox, S., 43
Fragmentation of chromosomes, 125
Frame shift, 120
Fungi, 69

Galileo, 235
Gametes, 64–66, 70
Gametophytes, 66, 70
Gene, 120, 123, 133
  dominant, 121, **122**
  duplication, 126
  flow, 133
  pool, 45, 126
  recessive, **122**
Genetic(s), 121–26, **122,** 133
  code, 119
  drift, 127, 133
Genotype, 123, 129
  change in, 124–26
Geological
  history, 53
  time scale, 35
Gerard, R., 244
Gigantopithecus, 203, *217*
Grant, V. and K., 156
Gunflint Formation, 54

Haemophilia, 123
Haldane, J. B. S., 129
Hamilton, S. D., 184–85
Hand axe culture, 202
Haploid, 63, **64,** 70, *189*
Haplodiploidy, 184–86, *189*
Hardy-Weinberg law, 127, 133
Hegel, 243, 250
Heliconius, **152**

Hemichordate, 98
Heraclitus, 1, **2**
Heterotroph, 70
Heterozygote, 122, 133, 151
Holism, 19, 26
Homeostasis, *189*
Hominid
  evolution, 206–16
  geological ranges, **202**
  phylogeny, **192**
  relationships, **196**
  shift in environment, 207
Homo, 189
  relationships, **196**
Homo erectus, 204–5, 211–12, *217*
Homo sapiens neanderthalensis, **198,** 205–6, *217*
Homo sapiens sapiens, 195, *217*
  evolution, 213–16
  fossil record, 206
  origin of races, **215**
  rate of evolution, 219
  skull, **198**
Homology, 137, 139, **140,** *168*
Homozygote, 122, 133, 151
Hormone, 189
Humanism, 26
Hume, D., 236
Hummingbirds, 152–54, **155,** 156–57
Huxley, J., 244–45
Huxley, T., 244
Hybridization, 152, *168*
  animals, 161–63
  DNA, 133
  plants, 160–61
  suture zones, **166**
Hymenoptera, 174–80
  castes, 175–80, 184–85
Hystrix, **141,** 142

Ideology, *255*
Immanence, 20–23
Indeterminacy, 9, 10
Industrial melanism, 128–29, 148–**150**
Insects
  and flowers, 154–55
  nests, **178**
  larval feeding, 158–59
  origin social groups, 183–85
  social groups, 174–87
  social hierarchies, 173
Interphase, 63, **64**
Introgression, *168*
Inversion of chromosomes, 124, **125**
Isolation, 111
  reproductive, 132

James, W., 1
Jellyfish, 102
Joad, C., 21

Kant, I., 1, 236–38
Kenyapithecus, 203
Kettlewell, H., 128, 149
Knowledge, 226–30
  applied, inferred, inspired, 231

Lamarck, 24, 238
Language, 231
Leppick, 154, **155**
Liberalism, 255
Lingula, 84
Life
  as immaterial force, 23
  as organization, 24
  cycle, **66–67**, 70
  immanent, 20–22
  nature of, 13–27
  origin of, 29–50
  philosophical points of view, 18–26
  reductionist points of view, 17
  scientific point of view, 15–18
Life-matter problems, 15–25
Lipids, 47
Logical positivism, 252, 255
Lycus mimetic complex, 147, **148**
Lyell, C., 238

Mammals, 91–95
  North and South American, 167(t)
  skulls of, **94**
Man
  culture, 202
  evolution, 171–218
  fossil record, 200–6
  jaw action, 197–98
  skull, **198**
  tooth row, 197
  what is man?, 195–200
Margulis, L., 59, 61–62
Marsupials, 93, **144, 145,** 146
Marx, K., 243, 248
Materialsim, 3, 6, 10
Mating patterns, 127, **128**
Maupertius, 237, 238
Mayr, E., 115
Mechanism, 10
Mechanistic outlook, 6
Meiosis, 52, 62–68, **64**, 70
Mendel, G., 109
Metabolism, 16, 47
Metaphysics, 255
Metaphyta, 51, 62, 70

Metazoa, 51, 56, 73–99
  origins, 75–83
  radiations, 83–95
  sources, 81–82
Micelles, 43–45
Michener, C., 173, 187
Mimicry, 146–48, 152
  Batesian, **106,** 146, *168*
  Müllerian, 147, **148,** 153, *168*
Mitochondria, 60, 61, 70
Mitosis, 62–68, **64,** 70
Mode of thought, today, 252
Molecules, 47
Mollusca, 98, **104, 105**
Monarch, butterfly, **106,** 146
Monera, 51, 52, 54
Monomeres, 39, 40–41, 47
Monophyly, 59
Monotremes, 93
Moon, capture of, 37
Moore, G., 244
Morals, 255
Mousterian, culture, 202
Mutations, 117, 119, **125,** 133
Mysticism, 255
Mythology, 255

Natural selection, 8, 10, 45, 116, 126–30, 132
Naturalism, modern, 252–53, 255
Nautilus, 86
Neanderthal man, **198,** 212–13
Neodarwinian selection, 29, 31, 45
Neolithic, 202
Nests, insects, **178**
Newton, I., 235–36
North and South America, 139–45
  mammals, **144**
Nucleic acid, 41
Nucleotide, 47, 117

Objectivity, 26
Olduvai, 210
Onychophora, 98
Oparin, A., 42
Oreopithecus, 203
Organic molecules, 39, 40
Origin of life
  creation, 33
  environment of, 36–38
  evolutionary model, 44–46
  extraterrestrial, 32
  spontaneous, 32
  time of, 34–35
Origin of species
  effects of, 239–56

stage for, 234-39
Owen, R., 241
Oxygen, in atmosphere, 79

Paleolithic, 202
Parallel evolution, 136, **137**
Parallelism, *168*
Paramenides, 1, **2**
Paranthropus, 192
Parazoa, 98
Parthenogenesis, *190*
Pebble tools, 202, 210
Phenotype, 45, 114, **123,** 129, 134
Pheronomes, 174, *190*
Philosophy, 10, 249-63
Phoronida, 98
Photosynthesis, 45, 52, 70
Phylogeny
    animals, **77**
    general, **74**
    hominids, **192**
    primates-perissodactyls, **199**
Phytoplankton, 79
Pithecanthropus, 204
Placental mammals, 93
Plant pollination, 153-54
    bats, 153
    hummingbirds, **157**
    insects, 153
Plastids, 60, 70
Plato, 1, **2**
Pleiotropy, 123, 134
Pogonophora, 98
Pollination, 153-54
Polygenic, 123, 134
Polymer, 39-41, 47
Polymerization, 31, 47
Polymorphism, 151-53, *168*
Polyploidy, 124, 134
Popper, K., 226, 229
Population, 110, 113, 115, 132
Positional changes in chromosomes, 124, **125**
Positivism, *255*
Pragmatism, 6, 10, 251
Prebiological evolution, 29
Precambrian, 53-56
    algae, **107**
Primates
    fossil history, 200-6
    phylogeny, **196, 199**
Primordial cell, 56
Probability, 6, 10
Procaryote, 30, 54, 56-58, **61**
Prophase, 63
Protective coloration, *169*

Protective color, form, 148-50
Protein, 48
    enzyme, **119**
Protenoids, 43, 45
Protista, 51, 71
Protostomia, **74,** 98
Protozoa, 98
Pseudocoelomates, 98
Purshia, **161**

Races of man, **215**-17, *231*
Radiata, **74,** 98
Ramapithecus, 192, 209, *217*
Ramirez, W., 154
Rationalism, 10
Raven, P., 158-59
Recombination, 121
Reductionism, 17, 26
Regressive evolution, 163, *169*
Relativism, 20
Relativity, 9
Remington, C., 162-63
Replication, 16
Reptiles
    origin, 89, **90**
    skulls and jaws, **92**
Ribosomes, 118, **119**
RNA, 117, 119
Robinet, 20
Rousseau, J., 236-37, 244

Salthe, S., 43
Schopf, J. W., 35
Science, 10
Selection, unit of, 116
Sentience, 26
Sexual reproduction, 67, 126
Sibling, *189*
Simians, 206
    habits, 207
Simpson, G., 115, 234, 244-45
Sinanthropus, 204
Sipunculoidea, 99
Skepticism, *255*
Smilodon, **145**
Social Darwinism, 247-49, *255*
Social systems, 171-89, *189*
    Evolution of, 182-84
    insects, 173 (t)
    vertebrates, 173 (t)
Socialism, 256
Soudan formation, 54
South American mammals, 143-45, **144,** 167 (t)
Speciation, 111-12
    allopatric, 112

Darwin's finches, **139**
  phylletic, *168*
  sympatric, 112
Species, 109–16, 132
  allopatric, biological, dynamic, temporal, sympatric, *132*
  nature of, 110–16
  origin of, 111, **112**
Spencer, H., 244, 248
Sperm, 120
Spinoza, 1, 20
Spirochaetes, **61**
Spore, 71
Sporophyte, 66
Stromatolites, 54
Stutz, H., 160
Subjective, 26
Suture zones, 163, **165,** *169*
Symbiosis, 59, 156, *169*
  in origin of eucaryotes, 61–62

Tardigrada, 99
Taylor, O., 162–63
Technology, *256*
Teilhard de Chardin, **14,** 22, 242, 250–51
Teleology, 5, 10
Termites, 185
  eusocial structure, 180
  nest, **178**
  origin of social castes, 181
Terrebratula, 84
Tetraploid, 124
Thallophytes, 52, 71
Theology, *256*
  and evolutionary thought, 240–42

Therapsids, 89, 91
Thermodynamics, 27
Thomas, L., 60
Thylacosmilus, **145**
Transcription, 118, 134
Transformation, 124, **125**
Translation, 118, *134*
Translocation, 124, **125,** *134*
Turner, J., 152
Typology, 132

Variation, 10, 116–26, 132
  chemical, 117–21
Vertebrates, 105
  radiation of, **89**
Viceroy butterfly, **106,** 146, 147
Viruses, 15
Vitalism, 6, 10, 18–19
Voltaire, F., 235–36

Wallace, H., **14**
Ward, W., 152
Wasps
  castes, **176**
  eusocial structure, **176**
Whitehead, A., 22–23, 250
Williams, G., 182, 186
Wilson, E. O., 172–89
Wormlike phyla, 99
Wright, S., 129

Zygote, 63, **64,** 65, 71